SCOTLAND'S PAST IN ACTION

Making Cars

Alastair Dodds

NATIONAL MUSEUMS OF SCOTLAND

Published by the National Museums of Scotland
Chambers Street, Edinburgh EH1 1JF

ISBN 0 948636 81 5

British Library Cataloguing in Publication Data

A catalogue record for this book is available from the British Library

Series editor Iseabail Macleod

Designed and produced by the Publications Office of the National Museums of Scotland

Printed on Huntsman Velvet 110 g²m by
Clifford Press Ltd, Coventry, Great Britain

Acknowledgements

The author is grateful to all those who helped in many ways, including: Alastair Smith and Susie Stirling, Museum of Transport, Glasgow; Mike Ward, Grampian Transport Museum; Brian Lambie, Bigger Museum Trust; James Savage, Edinburgh; Bob Allan, The Imp Club; Robert Grieves, Paisley; Neale Lawson, Dumfries; Lewis Martin, Jersey; Michael Mutch, Myreton Motor Museum; the staff of the libraries in Paisley, Dumfries, Dunfermline and Hamilton; and my wife Anne for her helpful comments and many hours of proof reading.

Illustrations: Front cover, back cover, 22, 28, 41, 48, 66, 74, 75: Glasgow Museums and Art Galleries. 4, 6, 7, 9, 17, 27, 36, i top, ii top and bottom, iii right and bottom, 62, 72, 78: NMS. 10: Trustees of the National Library of Scotland. 13: Dunfermline Library. 16, 30, i (bottom), iii top, 81: Grampian Transport Museum. 18, 21 top and bottom, 31: Albion Archive, Biggar. 25, 35: Dumfries Museum. 27, 33, 34, 76: Robert Grieves. iv (top): Argyll Turbo Cars Ltd. iv (bottom): Williams Grand Prix Engineering. 42: James Savage. 43: RA Skeoch. 47: National Motor Museum. 52, 53, 55, 57, 58, 59: The Imp Club. 68: Haldanes of Cathcart. 80: Scotsman Publications.

Front cover: *Illustration adapted from the 1906 booklet produced to commemorate the opening of the new Argyll factory at Alexandria.*

Back cover: *A detail from a Hillman brochure advertising the many accessories which were available for the Imp. The focus of attention here is the spotlamps.*

CONTENTS

1 The dawn of motoring

From the pneumatic tyre to the television, Scots have always been at the forefront of invention, particularly in the fields of engineering and science. It may be surprising that such a small country could have produced so many great inventors but, having accepted this, it is then no surprise to find Scots at the forefront of the development of road transport.

The world's first pneumatic tyre was patented in 1846 by Robert William Thompson of Stonehaven to improve travel in horse-drawn carriages. Another Scot, John Boyd Dunlop, reinvented the pneumatic tyre in 1888 and founded a company which became a household name synonymous with car tyres. Even the very roads upon which we drive would only be mud tracks were it not for a Scot, John Loudon MacAdam, who developed in the 1820s a hard surface using small stones. In time this had tar added to become the tarmac road of today.

The invention of the modern motor car is usually credited to a German, Karl Benz, in 1885, but the search for a replacement for the horse as the principal means of personal transport started long before then. One of the earliest known experiments in motorized transport was made by a Scottish engineer William Murdoch about 1784. While working for the famous partnership of Bolton and Watt, erecting pumping engines in Cornwall, Murdoch built a small three-wheeled carriage powered by a spirit-fired steam engine.

This small model, which measured only about 20 inches (50cm) in length, was very much based on the existing technology

The clock tower and carving above the main entrance to the Argyll works at Alexandria reflecting the aspirations of Scottish engineers of the nineteenth century.

A Scottish engineer, William Murdoch, designed and built what was probably the first motor vehicle in Britain. This replica of his model is in the collections of the National Museums of Scotland.

of the day, with the power from the single-cylinder engine driving a crank on the two rear wheels via a rocking beam. Although it looked like a tiny mine-pumping engine on wheels it established the principle of motorized transport. The model was demonstrated on a number of occasions, usually on a good, level, well-made piece of ground. On one particular evening it was sighted running along the church path by the vicar of Redruth, who believed that he had encountered the devil.

The story goes that Murdoch was on his way to London to patent his design when he was intercepted by Matthew Bolton who persuaded him to abandon his experiments. Probably neither Bolton nor his partner James Watt wished their valuable employee to waste his time on steam carriages. Whether a full-size version would have been successful will never be known, but given the state of the roads at that time it is unlikely. The original model is preserved in Birmingham Museums.

There is a possibility that Watt himself had an interest in building a steam-powered vehicle, as later in his life he wrote that his attention was first drawn to the subject by Dr Robison at Glasgow University, who had proposed applying a steam engine to drive a vehicle. Watt had even gone as far as to take out a patent protecting his ideas though he was far too busy with the stationary engine business to take the idea any further.

Although there were examples of steam carriages prior to 1784, notably that of Cugnot in France, it is almost certain that Murdoch's model was the first motorized vehicle to be built in Great Britain. It is interesting to speculate as to where Murdoch's experiments might have led had he not been persuaded to abandon them. He later became famous for discovering the use of coal-gas as a means of lighting homes and streets.

The first vehicle to be built in Scotland was almost certainly an experimental model by William Symington. Symington was the engineer to the lead-mining company at Wanlockhead, and was responsible for the mine pumping engines there. He is probably

William Symington built the first steam carriage model in Scotland and demonstrated it in Edinburgh in 1786 - truly a horseless carriage.

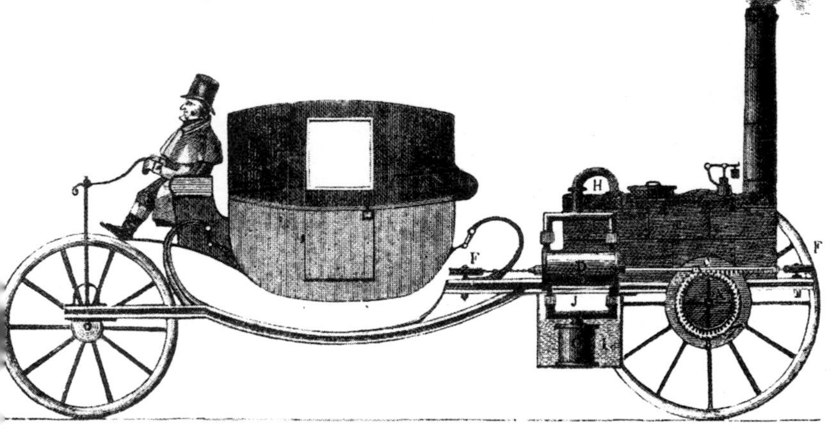

best known for his experiments into steam-powered ships and for the building of the first successful full-size steam vessel, the *Charlotte Dundas*.

An event which may have sparked William's interest in motor vehicles was a visit by Murdoch to Wanlockhead in 1779. Murdoch was there to commission a Watt engine which had been erected by George Symington with assistance from William, his younger brother. There is little doubt that William would have had plenty of opportunity to talk to Murdoch during this period. Symington's steam carriage model was exhibited in Edinburgh in 1786 and is reported to have been successful. It was certainly an ingenious design being driven by a pawl-and-ratchet. However, the state of the roads at that time, and the probable weight of a full-size carriage, deterred him from continuing his experiments. No doubt he also saw the application of steam to ships as a more practical use with the building at that time of a number of new canals. Further models were built in Scotland including one by Burstall and Hill of Leith in 1824. This took the form of a typical carriage of the day mounted in a chassis, with the boiler and engine behind. The design was a complicated one and again would probably not have translated well into a full-size vehicle.

Probably the first full-size Scottish-built steam carriages, and the first to carry fare-paying passengers, were built in Edinburgh about 1834. Operated by the Steam Carriage Company of Scotland between Glasgow and Paisley they ran successfully for a number of months until an accident caused one to overturn, landing on the boiler. The resulting explosion killed some of the passengers, and the service was ended. The carriages had been designed by John Scott Russell the naval architect who built the steamship *The Great Eastern*. Here again we find a link between ships and roads, though this is perhaps not very surprising with steam being the only power available and most of the advances and expertise in lightweight engines being for maritime use.

The use of steam power on the roads posed other problems apart from the dangers of the boiler bursting. Despite the advances in steam-engine design they were still heavy and required frequent stops to replenish the water supply. The fuel also was unsatisfactory, with coal or wood being heavy and inefficient. These fuels also required a stoker whose weight added to the overall inefficiency of these vehicles.

There were many partial successes including those of A W Forbes, an Edinburgh engineer, with a steam car in 1866. Another engineer, L J Todd of Leith, built a two-seat steam car in 1869 and a five-seat version in 1870. The most famous name involved in steam carriages at this period was Robert Thompson of Stonehaven, the first inventor of pneumatic tyres. Thompson's Road Steamers used solid rubber tyres and were sold in some numbers, some being built in Leith about 1867.

There were many other experiments taking place in Scotland including an interesting early use of electricity for a vehicle which was capable of carrying people. This was built by Robert Davidson of Aberdeen as a prototype for an electric railway carriage. The

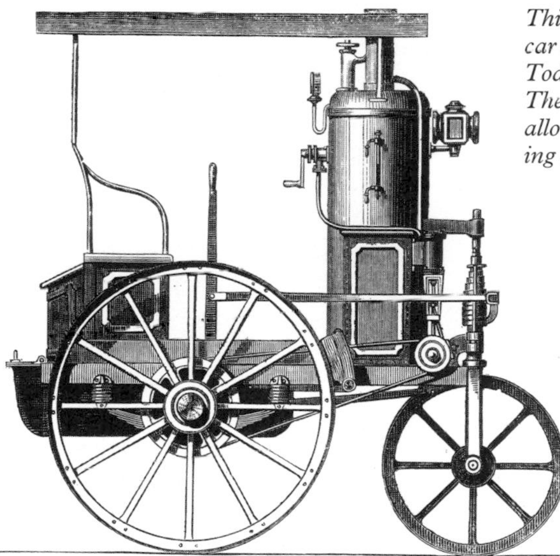

This two-seat steam car was built by L J Todd of Leith in 1869. The three wheels allowed a simple steering arrangement.

vehicle was capable of carrying two people across a wooden floor, and showed enough promise to attract funding for the project. Unfortunately when the resulting carriage was tested in 1842, on the Edinburgh and Glasgow Railway, it could only manage four mph. Nothing further was heard of this idea.

At the same time as the steam-powered monsters were being built in the late 1860s, a new industry was being founded in the English Midlands which would soon provide much of the man-power and expertise needed for a motor industry. This was the bicycle industry where highly skilled mechanics were building lightweight machines using new techniques. Cycle building soon spread to Scotland, with companies such as Victoria and Howe in Glasgow, and before long was feeding the new Scottish car industry.

The only component missing before a lightweight motor car could be built was the engine. The basis for an engine had been

David Dorward of Dumfries with his 1876 steam car. The engine was by the same firm who later built their own car, the Drummond.

around for some time in the form of gas engines but these were in the main large units comparable in size and weight to contemporary steam engines. The breakthrough came when a German engineer called Gottlieb Daimler, who was employed at the Otto gas engine works, developed a lightweight engine which used petroleum spirit vapour in place of gas. Daimler's engine was ready in about early 1885 and he first applied it to a vehicle in the form of a motorcycle in 1886. At the same time Karl Benz was also experimenting with the four-stroke petroleum engine. He built the three-wheel car which was to become the forerunner of all future development.

At this stage there was still no clear precedent for the number of wheels which should be used for cars. Many horse-drawn vehicles used four wheels, while the three wheels of the pedal tricycle were a common sight on the roads. Perhaps the engineering used in the construction of bicycles and tricycles was seen as a closer relation to the car than the crude horse-drawn carts. It is probably not surprising that Karl Benz employed a firm of cycle makers to build his first car. A more practical reason for using three-wheelers may have been the much simpler steering mechanism required. Even into the 1890s there were a number of makers persisting with three wheels, including some larger manufacturers such as Leon Bollée, in France.

Much of the early development in motorcars took place in Germany and France. In Great Britain there was no such development, resulting in the first cars being built using imported kits of parts. These were usually assembled using a locally coach-built body. One such firm was Arnolds, a firm of agricultural engineers from Kent. An example of one of their Arnold-Benz cars, dating from 1897, can be found in the National Museums of Scotland collections. This and similar enterprises provided useful experience to the founders of the emerging motor industry.

Scottish coach-builders quickly realized the sales potential of the new horseless carriage, possibly even realizing that horses were soon going to be replaced by the new invention. Among the earliest was John Stirling of Hamilton who acquired his first chassis kit from Daimler in 1896. By early in the following year he was buying the chassis in batches of 50 at a time. Most of these early cars were based on the coach bodies with which the Stirling family business were familiar. Dogcarts and wagonettes were popular but other styles such as Stanhopes and van bodies were available, all to the customer's individual order. French chassis such as Panhard and Clement were also used and at least three of the Stirling-Panhard cars still exist today.

The assembly of cars was not, however, confined to the West of Scotland. In Aberdeen the Caledonian Motor Car and Cycle Company of Union Street, was also building Daimler cars to order. Lighter De Dion engined voiturettes were also available which possibly used a chassis built by Caledonian themselves.

Not all vehicles built at this time were intended for series production. One of the most interesting individual machines was a three-wheel steam carriage known as the Craigievar Express. This was designed and built by Andrew Lawson, the postman in the Craigievar area of rural Aberdeenshire. Construction was started in 1895 and 'Postie' Lawson first drove it in June 1897. Powered by an engine bought second-hand from a sawmill in Aberdeen the vehicle had a top speed of 10mph(16Km/h). Lawson used the Express infrequently because, as he once said, 'if coals were cheaper I should have had it going a good deal'. The success of the vehicle can be gauged from the fact that it is still regularly steamed by the staff at Grampian Transport Museum where it is now preserved.

Lawson had no illusions about his vehicle being anything other than a home-built experiment. However, it is obvious that he realized that there was a future in motor transport. In a letter which he wrote to *The English Mechanic and World of Science* in 1901 about his 'steam motor-car' he describes it, and comments 'not that it will commend itself to your readers by any means. I have sent this as a sort of Clown to amuse the audience before the curtain rises.'

2 Founding an industry

The building of continental designed cars was proving to be a successful new industry in Britain. Although many of the imported components were modified and improved upon by British engineers, the time was ripe for an indigenous design. Possibly the first such design which owed nothing to continental cars was the English-built Lanchester of 1896. However, at the same time a Scottish engineer George Johnston was at work on his own design of car and had probably built a prototype during 1896.

Johnston was a locomotive engineer working in Glasgow who had imported a German-built Daimler in October 1895, the first car in Scotland, and had studied its design with interest. Being a practical engineer he noted a number of flaws in the design which he felt could be improved. His next car was a Panhard which, although a much better vehicle, still contained a number of deficiencies.

Johnston now went into partnership with his cousin, Norman Fulton, to develop a design of their own. They were joined by Thomas Blackwood Murray, an experienced engineer who had graduated from Edinburgh University as a Bachelor of Science in Engineering. He was later to use the experience gained at this time to start his own company. This small group started by experimenting with electric power but soon abandoned that line of development. They discovered, as earlier engineers had, that the

Designed by a Glasgow man and built in Dunfermline, Fife, the Tod three-wheeler of 1897 was contemporary with George Johnston's early experiments.

weight of the batteries required for even relatively short journeys made the vehicle too heavy to carry any significant payload. Only in the USA did further development continue.

Although a number of steam cars were still being built in France and the USA at this time, and by isolated makers such as Simpson in Scotland, it was generally accepted that the internal combustion engine was the way to proceed. With this in mind the trio set out to build their own new design with a view to putting a car into production.

To finance the venture a company was formed and named the Mo-Car Syndicate Limited with an address at Bluevale, Camlachie in Glasgow. The first major backer to be found was Sir William Arrol, a civil engineer whose greatest achievement at that time had been the construction of the Forth (Rail) Bridge. Another backer was Archibald Coates whose family had interests in the textile industry and produced the famous Coates threads.

The company was formally constituted at the end of 1895 with Arrol as Chairman, Johnston as Managing Director, Fulton as Works Manager and Blackwood Murray as Commercial Manager. This was the first company established to manufacture cars exclusively designed and built in Scotland. The new car was named Arrol-Johnston after the financier and the founder. The prototype petrol-engined car was soon on the road but it was not until 1899 that production started. It took the form of a six-seater dogcart fitted with a two-cylinder engine driving traditional wooden wheels with solid tyres through a chain drive. The only concession to developments taking place on the Continent was the use of a steering wheel in place of a steering tiller.

This first period of the company was to be short lived. After only a handful of cars had been built, there was a disastrous fire early in 1901. Everything was lost including all the company records and drawings. Although the company was in production again within a year the loss of the historical records of these early years, of what was to become Scotland's largest car builder, has grieved historians ever since. While the prototype Arrol-Johnston is claimed by some to have precedence over the Lanchester as the first indigenous British car, an equally important first can be claimed for Edinburgh in having the first purpose-built car factory in Britain.

Although some of the early experiments with steam-driven vehicles had taken place in Edinburgh it was not until the founding of the Madelvic Carriage Company in January 1898 that motor car production came to the city. More precisely it was to be, literally, a motorized carriage. The company was founded by William Peck (later Sir William) who was the Edinburgh City Astronomer and lived in the observatory house on Calton Hill. His idea was to add a motor to a standard design of carriage, presumably to cater for those who were unsure of the idea of a completely horseless carriage. Certainly the idea must have appealed to the many coachmen who at the time were not only having to learn the new business of being a chauffeur, but were in most cases acting as mechanics also. Breakdowns were very common with these early cars and the only way to get home was to get out and make a repair at the roadside.

Not only was the Madelvic novel in having a fifth wheel to provide the drive but it was also electrically powered by what were known at the time as storage batteries. In keeping with the rest of the design based on a carriage, and to cover the possibility of breakdowns, the vehicle was provided with means of fitting shafts for a horse. While the horseless carriage design may have been backward looking, the new factory was quite the opposite. The office was built cheaply in brick edged with stone, and featured a central carving depicting the Madelvic's fifth wheel with chain drive up to the motor. The office itself was not unusual but the main production block was, in so much as it was completely right for car production. A two-storey brick-built block utilized steel floor supports giving good clear working areas. An adjoining single-storey block was designed to give excellent natural lighting with a glazed roof. Another interesting feature of the site was the inclusion of a test track.

Typical of the period was this Mowat car built in Aberdeen about 1899. Mowat was also associated with the Harper car of Aberdeen.

The factory may have been designed well but it was not used properly. A combination of a poor flow of assembly through the building, a poor product, and a huge initial investment ensured that the project was doomed from the start. Certainly Peck must have realized his mistake in trying to use electric power because he went on to patent a number of designs for internal-combustion engines. Within two years the business was bankrupt, and the first British purpose-built car factory was for sale. It was bought by the Kingsburgh Motor Construction Company for £13,000; a bargain compared to the £33,000 which the works had cost to build. Kingsburgh did not last long either at Granton, as we shall see later.

The last year of the nineteenth century was to see the formation of two companies, each of which was destined in its own way to become a major player in the Scottish motor industry. The first of these was founded by Alexander Govan, an experienced engineer in his mid thirties, who had a small company assembling French cars in Hozier Street, Bridgeton, Glasgow. Unlike other companies involved in assembling foreign cars, such as Caledonian in Aberdeen, Govan had ambitions to build his own cars. Like George Johnston he thought that he could produce a better design.

Govan had excellent experience both from his training in Coventry and from the assembly of several different makes of car. He also had excellent premises which had previously formed the works of the Scottish Cycle Manufacturing Company. As so often in the early days of the car industry we again find connections with the cycle industry. It was with this background that Govan built his first car during 1899. Based on the design of the Renaults which he had been assembling, it was sufficiently successful to attract external financing for series production.

The original Argyll works in Hozier Street, Bridgeton, Glasgow.

The chaotic interior of the first Albion factory at Finnieston Street, Glasgow.

Funding came from Walter Smith of the National Telephone Company, who put up £15,000 to start a limited company called Hozier Engineering. The first cars were built in April 1900 and were known as Argylls.

The other significant company to be founded in 1899 was Albion which was also Glasgow-based. Both men behind this venture had been involved in George Johnston's Mo-Car Syndicate building Arrol-Johnston cars. One was Norman Fulton, Johnston's cousin, who had left Mo-Car during 1899 to make a study tour of car production in America. The other was Thomas Blackwood Murray. It was during his time as works manager at Mo-Car that Murray had begun to design his own car, and when Fulton returned from America they were ready to build a prototype. Their funding came from Murray's father John who mortgaged his farm in Biggar, Lanarkshire for £1500. Although a far cry from the £15,000 raised by Govan it was enough to move into premises shared with the Clan Line Repair Shop at Finnieston Street, Glasgow, and buy the materials for their first car.

The scene was now set for the rapid expansion of the new motor industry in a new century. Many of the small pioneering companies such as Simpson, Stirling and Caledonian, building cars using old technology, would fail over the coming few years. But many others would shortly be formed to feed the growing enthusiasm for the car as a means of transport.

3 A flourishing industry: the car comes of age

Nineteen hundred was a significant year: not only was it the start of a new century but it also ushered in a new age, with new attitudes to transport. No longer was the car seen purely as a novelty with little practical use, but was now embraced by a growing band of enthusiasts who saw a future in personal transport in which the horse played no part.

The bicycle had played a leading role in changing peoples attitudes, with the formation of clubs to promote the use of cycles and to lobby for better roads. The car enthusiasts now began to band together and take over what the cyclists had started. Restrictive legislation had held back motoring in the UK despite the repeal, in November 1896, of the famous 'red flag law' prior to which cars were restricted to four mph and had to be preceded by a man with a flag. An Emancipation Run from London to Brighton was organized in 1896 by the Motor Car Club to celebrate the new freedom.

Now in the twentieth century the time had come for rapid advances. This was particularly notable in the technology of the motor car. The half dozen or so small companies building cars in Scotland in 1900 were for the most part using old designs and had little capacity for introducing changes. However, the three newly formed companies of Albion, Argyll and Arrol-Johnston were soon to dominate the Scottish car industry, using better metals and techniques to overcome the reliability problems suffered by the industry in general.

Of the three new companies it was perhaps Argyll which produced the most up-to-date design. The car featured a forward-mounted engine which, although not altogether obvious at that time, was soon to be the conventional position. The drive from the engine to the rear wheels was also novel in having shaft drive at a time when chain drive was more usual. There were areas of the design which still owed more to the cycle industry including handlebar steering and the tubular chassis. This was not surprising with the equipment available in the former cycle works which Argyll occupied.

Arrol-Johnston and Albion produced an altogether different style of car, though each had nonetheless incorporated design features which were progressive. A feature shared by both makes which was far from progressive was the bodywork based on the old horse-drawn Dogcart. These were designed to carry hunting dogs under the seats in a compartment with louvres along the sides for ventilation. This design lent itself to the motor car with the engine filling the under-seat compartment almost directly above the rear wheels, which gave a suitably short run for the driving chains. It is not surprising that both these firms produced similar vehicles, with Murray and Fulton having worked on the design of the Arrol-Johnston before founding Albion.

This first Albion still survives, albeit without bodywork, in the Museum of Transport in Glasgow. Looking at the chassis it is possible to see how the Albion uses channel-section steel in place of the large diameter cycle tubing more common with earlier designs. Tubing was still used for bracing but the soon to be familiar chassis form was there from the start. The second car also survives, at the National Museums of Scotland in Edinburgh, and shows the wooden bodywork with slatted sides. Also to be seen is the poor seating arrangement, taken straight from the horse-drawn vehicle, with the rear passengers facing backwards with their backs to the driver and front passenger.

Albion dogcarts outside the home of John L Blackwood Murray at Heavyside, Biggar. The front car, driven by John Lawson the chauffeur, is the second car to be built and is preserved in the National Museums of Scotland.

The chassis and engine of the first Albion to be built. This can be seen in the Museum of Transport, Glasgow.

Perhaps even more bizarre was the arrangement adopted for many of the Arrol-Johnston cars where an additional two seats were positioned in front of the driver. This together with an already heavy body, many built in mahogany by the Glasgow firm of cabinetmakers Wylie and Lochhead, created an imposing vehicle when compared to the lightweight Argyll, and the neat Albion. Despite the bulk of the vehicle it produced a good performance for the period, with a top speed of 17mph (27km/h) on the level from the 12hp engine. Actually driving these early cars was a real challenge. One pioneer Scottish motorist, Mr G H Christie, provides a vivid description of driving his Arrol-Johnston as follows, 'the whole outfit was ideal to gather speed rapidly when coasting downhill. If a rut or a greasy surface from water or fallen leaves was encountered, the car would just not keep on the road.

This Arrol-Johnston dogcart was one of three which drove from Land's End to John O'Groats in September 1903.

Often we found ourselves in ditches or locked in hedges. We had a terrier who thoroughly enjoyed the thrill of these coastings, barking all the way until he was shot over the fence or hedge at final impact.'

These new cars were soon being given opportunities to prove themselves as the eager new owners put them through their paces in club outings. One of the first of these was organized by the newly-formed Scottish Automobile Club in the summer of 1900. Members from Glasgow and Edinburgh drove to Stirling where they joined up for a procession at the Highland and Agricultural Society show. Several Scottish-built cars attended including an Argyll driven by Alexander Govan, a Stirling driven by John Stirling, another Stirling and an Arrol-Johnston. The cars driven by Govan and Stirling were there as privately owned cars but it is obvious that the companies saw these events as an excellent chance to show off the reliability of their cars to other enthusiasts.

The run was a success with the only difficulties being caused by the dust stirred up by the cars, a recurring problem for early motorists. This club run was the precursor of many more organized by the growing band of enthusiasts. The run highlighted a need to demonstrate to the public the practicality of the car as a means of transport. The reliability of the home produced product also needed promoting in direct comparison to the many continental cars being imported.

The opportunity presented itself the following year when the Automobile Club of Great Britain and Ireland organized the first official car trial in Scotland to coincide with the 1901 Glasgow Exhibition. The first day of the trial covered 116 miles from Glasgow to Edinburgh and back. Scottish makers were represented among the 40 cars entered, with an Arrol-Johnston dogcart and a 4½hp Stirling. The starting point was the Corporation stables in Kelvinhaugh Street, Glasgow, and a reporter who was present described the scene: '...yesterday morning, Kelvinhaugh Street presented an unfamiliar appearance as the crowd of low-built motor cars manoeuvred about preparatory to the start. The interior of the storehouse, as they moved out one by one, resounded with the whir of their machinery. They were arranged in line along the side of the street, and at eight o'clock began to move off in close order through the city on the way to Edinburgh.'

The same reporter travelled with one of the cars and reported that 'There are few modes of travelling more exhilarating than by motor car'; but in the same report he states that 'There is the incidental discomfort of dust.' The second day of the trial was a route from Glasgow to Ayr and back, while the third day saw them visiting Callander. There were several breakdowns; the fate of the Scottish cars is not recorded, but in the main the trial was successful in promoting the car as a serious means of transport. Certainly there were large numbers of people lining the routes giving encouragement.

Meanwhile the industry in Scotland reflected what was happening in other countries with many small companies trying to cash in on the increasing demand for cars. Many of these firms came and went, hardly leaving any trace of their existence. William Robertson and Son in Dundee started building motor tricycles and light cars in 1900 but nothing further was heard of them after 1902. Kingsburgh, whom we heard of earlier, had taken over the Madelvic works at Granton in Edinburgh in 1900 and built a few 12hp cars before they also went into liquidation in 1902. The unlucky Granton works then passed to Stirling's of Hamilton who had some success building buses and lorries.

More than twenty car companies formed in Scotland after 1900 had vanished by the start of World War I in 1914. Many of the firms which started with high hopes only managed to produce single examples. Most went bankrupt in the process, or returned to their previous business having discovered that building cars was not as easy as they thought as designs rapidly became more complex. An example of this was the Edinburgh car dealer Alexanders who built a solitary 14/18hp Alex in 1908 before returning to selling other makers' cars.

An example of a firm which made a serious effort to move into car production was the Dumfries Brass and Iron Foundry in Dumfries. In this case over 125 cars of various types were built between 1905 and 1909 when the company went bankrupt. The Drummond car, as it was known, was the brainchild of an engineer and partner in the foundry, David Drummond. A hint as to why the firm failed can be gleaned from a reporter writing in *The Motor World* after visiting the works. 'What first struck one in looking round was the patent fact that the business must have developed too rapidly for the available space, the arrangement suffering as a consequence. The machine shop especially showed that the tools had been added as required, so that order could not be considered.'

The three big companies also had their share of difficulties with perhaps none as devastating as the fire at the Arrol-Johnston

The Drummond works in Dumfries which was described by a contemporary writer as being disorderly.

works, mentioned earlier. By 1905 the company was losing money and the original Mo-Car Syndicate was taken over by William Beardmore (later Sir William) who became the first chairman of the New Arrol-Johnston Car Company Limited as it was now known. Another director was the newly appointed chief engineer John Napier. Under Napier's guidance the design of the cars was updated and the new 18hp model was of conventional layout with a bonnet and radiator at the front, and a steering column and wheel. The day of the dogcart was finally over, but the company was still losing money and would continue to do so for some time.

One of the more radical moves by Napier after joining Arrol-Johnston was to enter the cars for racing. In 1905 the first TT (Tourist Trophy) races for cars were held in the Isle of Man, two years earlier than the first motorcycle event. Napier entered two cars, driving one himself. The roads were appalling and of the 54 starters only 24 completed the distance. But it was Napier who won at an average speed of nearly 34mph. The other Arrol-Johnston was fourth. The event was concerned with fuel consumption as well as speed, with both cars performing well. This was, however, to be the only success for Arrol-Johnston with the following five years all resulting in disaster through mechanical failures.

Argyll were also involved in competition and the same 1905 TT race which was won by Napier saw an Argyll come eighth; a good result considering the high rate of retirals. In March of the previous year Argyll had taken the Land's End to John O'Groats record, improving on the old record by ten hours. Six months later they improved on their own record, bringing the time down to an amazing 42 hours 5 minutes, particularly good considering that Highland roads were little more than dirt tracks in those days. Records such as this did a lot to convince the public of the reliability of cars in general and certainly helped sales of Argylls.

By 1905 Argyll were showing a healthy profit and Alexander Govan chose this time to restructure. The name Hozier Engineering went, and the company became Argyll Motors

Limited with the massive capital of £500,000. At this time it was decided that they had outgrown the old factory at Hozier Street, Bridgeton, Glasgow and they moved to a greenfield site at Alexandria, north of Glasgow. This new factory was to be one of the grandest ever built and was ultimately to be a major factor in the demise of the company.

The factory was opened on 19 June 1906 by Lord Montagu (father of the founder of the National Motor Museum at Beaulieu) at a cost of £220,000. The factory was built on a 25acre site adjoining a railway line, and just to the north of the Firth of Clyde, giving excellent communications for the distribution of the potential 2,500 cars a year which it was capable of producing. Unfortunately much of the cost of the factory was wasted in the construction of a 760ft (232m) long facade consisting mainly of offices. The centre of the frontage was surmounted by a huge clock tower beneath which a grand entrance hall and stairway,

The palatial frontage of the new Argyll factory at Alexandria, shortly after completion in 1906.

The spacious interior of the Argyll factory about 1906.

built entirely in marble, gave the impression of a palace rather than a car factory. Even the pediment above the main entrance was carved with a group of classical figures surrounding the nose of a motor car. Today the factory behind the facade has been demolished and the offices stand empty and vandalized awaiting a new use. Driving past this magnificent building it is difficult to comprehend that such a statement of success and confidence would see use as a car factory for less than eight years.

4 An industry in trouble

Only a year after the opening of the new Argyll factory Alexander Govan, at the age of only 38, died from a stroke, robbing the company of its founder and driving force. W A Smith remained as chairman but he had neither the flair nor the mechanical understanding that Govan had. A further strain on resources was caused by production continuing at Hozier Street despite the fact that the new factory at Alexandria was working at probably only

50% of capacity. The company was now in deep financial trouble and, despite a wide range of excellent cars on offer, the poor management and severe strain on resources caused it to seek voluntary liquidation in August 1908.

Over 1000 people lost their jobs, illustrating the importance of the success or failure of the larger Scottish car firms on the local economy. The product was a good one with a ready market for quality cars, so a new issue of capital was sought and the company reformed as Argylls Limited. After debts had been settled and assets written off the new company found itself with £100,000 capital. The wide range of cars on offer was cut from seven different chassis to only three. The ridiculous situation where 29 body styles had been available was rectified with a dramatic reduction.

Albion meanwhile was suffering none of the difficulties of its two main Scottish rivals. In 1903 the firm moved out of the old Finnieston Street works to a brand new factory at Scotstoun on the banks of the River Clyde. The factory, unlike Argyll's later monstrosity, was of a modest size with few pretensions. This was to place the firm in a good position for a steady growth over the coming years. From a total staff of 77 in 1903 there was a steady increase to 283 in 1906, with an equally steady rise in output from 33 cars in 1903 to 221 in 1906. The range of cars was kept simple with the 16hp A3 model replacing the 12hp in 1904 and this being joined by a 24hp car in 1906. That year also saw further ground being acquired adjacent to the factory for further expansion.

Unlike its rivals, Albion did not enter motor sport events but concentrated instead on reliability trials, winning a silver medal in the 1903 Scottish event and a gold in 1905. No doubt the patent lubricator and engine governor, both designed by Blackwood Murray and used for over 20 years, contributed to the reliability of the cars and the excellent reputation which the company gained. Although much of the early reputation was from the cars, production of commercial vehicles was increasing, with a general move away from simply putting commercial bodies on car chassis, to heavier specially designed vehicles.

A typical Highland road in the early years of this century. An Albion landaulet stuck in mud near Braemar.

By 1913 the company was employing nearly 1000 people, and the anual output had increased to 554 vehicles, though only 125 of these were private cars. Car manufacture continued until the end of 1913 when the directors decided to concentrate solely on commercial vehicles. A significant expansion of the factory was completed just in time for the company to be able to make a major contribution to the mechanization of the British army during World War I. From there Albion went on to become a world-class builder of trucks and, after several takeovers, still exists today as a company producing back axles for trucks. While Albion was going from strength to strength Arrol-Johnston's troubles continued. The firm's original financier, Sir William Arrol, had left the company at the time of Beardmore's takeover in 1905 having received £18,000 for his share of the company. George Johnston, its founder, continued with the firm, but he was dissatisfied with the way things were going and left to form the All-British Car Company in the old Greenhead Weaving factory at Bridgeton, near to the first Argyll factory. Despite grand plans there were probably no more than about twelve of the new eight-cylinder ABC produced. Within a year the firm had folded and Johnston left the motor industry to start a new business with a coal-refining plant in Glasgow, before eventually moving to America.

Arrol-Johnston badly needed new leadership if it was to survive. In April 1909 this was found when Thomas C Pullinger was brought in as general manager. Here was a man who knew the industry well, having trained in the cycle trade and worked for Darracq in France before moving back to England to design cars for Sunbeam and latterly Beeston Humber. William Beardmore himself had approached Pullinger with an incentive offer of one eighth of the company shares, to be paid for out of the profits. Under his leadership the company's fortunes changed and the next four years were to prove to be its best.

The same month in which Pullinger joined the company Ernest Shackleton returned from a South Polar Expedition amid great publicity, not least for having used a motor car built by Arrol-Johnston. The car was a special incorporating an air-cooled engine by Simms to avoid problems with frozen coolant. It resembled a pick-up truck and had skis fitted to the front wheels, with deep ribbing on the rear wheels for grip in the snow. Other ingenious features included using the exhaust to heat the carburettor and provide foot-warmers for the crew. The design was a complete success despite the company's lack of expertise in polar transport.

Pullinger's first move was to introduce an entirely new model of car which was ready for the 1909 Motor Show. Although generally of conventional design the car used a fluted bonnet with the radiator behind the engine in the style of the Renault of the same period. The car bore the slightly unusual designation of 15.9hp, with the fraction working as a positive publicity feature to make the car stand out as being new. The 15.9 was soon being produced at a rate of over 300 per year and was to become the company's staple product. The .9hp became an Arrol-Johnston feature with a six-cylinder 23.9hp and a smaller 11.9hp being introduced to the range for 1911.

Throughout the Edwardian period travel by motor car was still an adventure with many chauffeur/mechanics still being employed to operate these labour-intensive vehicles. It was not until 1913 that there was a general move towards fitting electric starters to cars, and even then usually on the more expensive models. Even long after the electric starter had become the norm most firms still fitted a hand crank to augment the generally unreliable batteries and electrics. Even as late as the 1960s the Scottish-built Hillman Imp was designed so that the wheel brace could be used as a hand starter.

Another feature of cars prior to World War I, which would have to be changed before widespread acceptance as a principle means of transport, was the acetylene lighting. This involved filling a chamber with calcium carbide which, when mixed with water, produced acetylene gas. This was then piped to the lights, each of which had to be lit in turn. After use the carbide had to be scraped out and replenished. Oil lamps which were just as messy, and produced less light, were the only alternative. Arrol-Johnston were one company which realized that the way forward was to produce an inexpensive car with both electric lighting and starting, offering this on the 15.9hp in 1914 at only £360.

With general improvements in car design the market began to expand in the years leading up to World War I. The Scottish makers had stiff competition from England and abroad, notably France and the USA. The ubiquitous Ford Model T was appearing in large numbers, being sold through dealers such as Alexanders in Edinburgh who still specialize in Ford today. One of the more bizarre publicity stunts of the period was undertaken by Henry Alexander, son of the founder, who made the first ascent and descent of Britain's highest mountain in 1911. Although there was a track of sorts up Ben Nevis 4406ft (1343m), which had served the observatory on the summit, this was still a considerable feat and must have helped sales significantly. It is interesting to note that while production of the Model T ran to a total of fifteen million over 20 years, the total output of the Scottish industry prior to 1930 was probably no more than 25,000 private cars.

In the remaining years leading to the outbreak of war Argyll struggled on, even managing to increase production to 450 cars in 1910. This was a fraction of the potential output from the new Alexandria factory, which by 1914 was costing £12,000 a month to run. New models had been introduced and the Argyll cars were well-built, stylish cars. However, no amount of reorganization

Punctures were a common problem for the early motorists. Here Mr Henry Archibald of Wishaw removes the wheel of his Argyll, while his family look on, about 1912.

was going to make up for poor management and huge debts incurred from the new factory.

The final straw came when Argyll lost £50,000 defending a legal action. In 1912 the company had entered into a licence agreement to build engines using the Burt-McCollum single sleeve-valve engine. This simple design, where the cylinder acted as a valve, was quiet and smooth compared to the standard poppet-valve used in most car engines. The design was successful, and enhanced the Argyll reputation for quality. The problem arose when Charles Knight, whose double sleeve-valve design had been turned down by Argyll, sued for infringement of patents. Despite successfully defending the action the financial strain was too great and Argyll went into liquidation once more in June 1914. In December the great Alexandria factory was bought by the Admiralty for the production of munitions. John Brimlow took over the company name with the Bridgeton factory the following year but only a few hundred more cars were ever built.

The old Arrol-Johnston factory at Paisley shortly before the company moved to Dumfries.

In contrast to the sad end to Argyll, which had greater potential than perhaps any other Scottish firm, the fortunes of Arrol-Johnston had never seemed better. Pullinger had pulled the company back from the brink of bankruptcy to the extent that a larger factory was needed. A greenfield site was chosen at Heathhall near Dumfries which offered excellent access to the English market, yet retaining the Scottish identity which was seen as a valuable asset.

Pullinger wrote to Henry Ford in America who, as Pullinger's daughter Dorothée later wrote 'was doing very well and building a fine factory', and was invited to make a study tour of the Ford plants. As a direct result it was decided that the new factory would be structurally to the same design as the Ford factories, as patented by Albert Kahn, which used concrete framing to provide a multi-storey building with large areas of glass providing natural light. The concrete frame, steel-framed windows and steel-clad doors gave an

Wooden car bodies in the paint shop of the new Arrol-Johnston factory at Heathhall. The clean, well-lit space is evident in this view.

additional benefit of a low fire risk, possibly prompted by the memory of the loss of the original works in Glasgow.

Although the site was rural there were few disadvantages. A siding from the Caledonian railway was brought right into the factory complex and the proximity to the border with England ensured excellent communications, both for raw materials and the finished product. Workers houses were built adjacent to the factory, and the town of Dumfries with a sizeable population, was only two miles away. A special passenger station was constructed adjacent to the works. Another factor in the choice of location may have been that the principal coachbuilder used by Arrol-Johnston was Penman of Dumfries. The prospect had never been so good for a Scottish firm to enter the major league of car production.

The new factory was started in March 1912 and opened by the Marquis of Graham in July 1913. The cost of only £85,000 seems good value when compared with the £220,000 which the Argyll

Characters by Car—No. 1

The Crofter

THE Crofter digs his meagre livelihood from the rocky earth amid the grandest, most rugged mountain scenery in the British Isles. But no man may dare the rough roads of these regions in anything but a reliable car.

The Arrol-Johnston 15.9 Touring Model (Type B. Chassis) is absolutely ideal. It is built for strain as well as style. The petrol consumption is ***20/25 m.p.g.,*** and the engine is built by the makers of the Beardmore Aero Engine—a fact that speaks for itself. A mechanical efficiency and service guarantee is given to cover the current year.

The Arrol-Johnston 15.9 Touring Model is the perfect car for the Owner-driver.

PRICE - - £650—with Full Equipment.

London Dealers :

LEVERETT, THORPE & KEARTON, Ltd., 122 New Bond Street, W. 1

ARROL-JOHNSTON, LTD., DUMFRIES, SCOTLAND.

works cost. In fact it was everything which the Argyll works was not, being in an excellent location and based on the very latest American working practices. There was even room for expansion on the 160 acre site.

When war was declared in August 1914, Arrol-Johnston found themselves in a very favourable position thanks to Thomas Pullinger who understood the value of diversification. He had been investigating the possibility of aero-engine manufacture since joining the company and had designed a crude four-cylinder engine himself, but the real possibilities came as a result of a visit to the Austro-Daimler factory in Vienna-Neustadt. With Beardmore's agreement he returned with blueprints and UK manufacturing rights, said to have cost only £10,000.

One of the first jobs was to translate information on the drawings from German into English. Pullinger's daughter Dorothée was working in the drawing office at the time and was one of those entrusted with this important work. Production was about to start when war broke out and there was great panic in the company because the first six crankshafts had been manufactured in Vienna and were being transported to Scotland by road. Pullinger himself travelled to Switzerland to see if the crankshafts had crossed the border and finding that they had, brought them back to Heathhall. Altogether over 2,500 of these engines were built. This fortuitous move into aero-engines was to ensure that Arrol-Johnston survived the war, in profit, and was in a strong position when car production resumed postwar. The company even undertook production of complete aircraft with 50 Sopwith Camels being built at Heathhall.

The engine became one of the standard engines used by the Royal Flying Corps and as the war progressed more power was required. Major Frank Halford was lent to Beardmore by the War

One of a series of adverts in 1921 which capitalized on the wartime aero-engine work at Heathhall to promote the reliability of the cars.

Office to help redesign the engine, and development work was entrusted to Pullinger. As a result the new engine was named the BHP, standing for Beardmore, Halford, Pullinger. A new company called Galloway Engineering was formed to build the engine and, due to the volume of work at Heathhall, a second factory was built at Tongland, near Kirkcudbright, some 40 miles west of Dumfries. Along with the expertise built up in the local workforce it was to prove to be a valuable asset for the company later on when they introduced a new marque of car in 1920, and needed premises in which to build it.

The war years were not always just hard work at Arrol-Johnston. Workers formed an entertainment group called the Auto-Knuts to entertain their colleagues. Performances were also given all over Dumfriesshire to raise money for benevolent causes. One show which they performed at New Abbey in February 1916 was reported by the local newspaper to have been a great success.

5 The industry flounders

In the aftermath of the Great War society had changed greatly, not least in the workplace. Women found a new freedom to pursue careers, soldiers returned to find their old jobs gone, and many companies which had turned to armament production were out of business. Perhaps the greatest change was to people's expectations, having fought and won a war which they believed would ensure that the country could remain free to prosper.

Something else had changed which was to affect the way people viewed transportation. Prior to the war motoring had been growing in popularity but was essentially still seen as the provence of enthusiasts and of the rich. However, large numbers of ordinary people had been trained to drive lorries, vans and cars in military service. Included in this new group were a number of women who had driven ambulances. Many of these people came home to civilian life with the desire to own a car of their own, and many of those with a job had money to spare.

The Scottish car industry should have been poised to take advantage of this new market by producing cheap small cars, but it did not respond. Albion had ceased car production in 1913 to concentrate on commercial vehicles and had become too successful in this area to go back to motor cars. Argyll had gone into liquidation in June 1914, and although John Brimlow had reformed the company in the old Bridgeton works it was a mere shadow of itsformer glory in the days of Alex Govan.

Apart from Argyll, the only Scottish car firm still in business in 1918 was Arrol-Johnston. By now well established in the Heathhall factory near Dumfries, the company were in reasonable financial health and had a relatively good design of car from immediately prior to the war which included electric lights and starting motor. A more sensible decision in retrospect might have been to update the pre-war 15.9hp and put it straight into production as a money making stop-gap until a new design could be produced.

Pullinger, however, was keen to be among the first to produce an a completely new car with which to quickly establish the Arrol-Johnston as one of the foremost cars in post-war Britain.To design the new car, he had persuaded G W A Brown to move to Arrol-Johnston from Clement-Talbot. Brown was a very experienced engineer, with wartime experience in the aviation industry, who quickly produced a design with all the latest features. A major advance was the centrally mounted gear and handbrake levers; pre-war it had been normal to mount these on the drivers right side, making it impossible to fit a driver's door.

Announced in 1919 as the Victory, the car had elegant styling and a 13hp engine in what proved to be a very light body weighing only a ton. The car was not the cheap, light-car which was needed for the new mass market, but it should have sold in reasonable numbers to the middle-class buyers. In the rush to get it into production it was not properly tested, and it turned out to be hopelessly unreliable. The car had potential if the problems had been sorted but its fate was sealed, ironically, by a publicity exercise which went badly wrong.

One of the first cars to be built, and the only one to be delivered, was supplied to the Prince of Wales for use during a holiday in Cornwall. The car broke down so often during delivery from a London railway station that it was loaded into a closed wagon at Exeter station late at night and returned to the factory. The publicity was so bad that no further Victory cars were built.

An updated 15.9hp car, with the radiator moved to the front of the car from the pre-war position behind the engine, took the place of the Victory and kept the company active. Perhaps a lesson was learned from this expensive mistake because Pullinger's next venture was to build a light-car based on the successful Fiat 501. This was to be built by a separate company under the trade name of Galloway, set up no doubt to segregate light-car production from Arrol-Johnston which had a reputation, albeit now tarnished, as a builder of quality middle-class cars. The new car was to be made in the detached factory at Tongland, near Kirkcubright, built during the war for aero-engine production.

The site at Tongland had originally been chosen because there was an untapped source of labour in the area which was not already engaged in war work. A large proportion of the workforce had been women and they were employed for the new enterprise. The managing director was also a woman, Pullinger's daughter Dorothée who had spent part of the war working for Vickers Ltd at Barrow, Cumbria. With experience in the Arrol-Johnston drawing office, and latterly as forewoman of the foundry core shop, Dorothée was an excellent choice. She brought to the job a great deal of enthusiasm and involved herself directly in the selling of the cars. On one particular sales trip to England, Dorothée made visits to Chelmsford, Bournmouth, Taunton and Torquay, illustrating the wide area targeted by the company.

Another benefit of the location of the factory was the cheap source of power from the River Dee. A report in *The Autocar* magazine of October 1920 states 'it is a wonderful setting for a modern factory building, which, approached from a tall stone bridge built centuries ago by sturdy monks, steals its power from

The Galloway factory with the tall netting round the rooftop tennis courts.

the river by trapping the troubled water in concrete pits containing slow speed turbines.'

Despite, or perhaps because of, the rural location of the factory the recreation facilities provided by the company were good. A bathing and swimming pool had been specially constructed and there were two tennis courts on the roof with a tall fence round the edge. There was also a very active hockey team; perhaps the only car factory to have had a hockey team but no football team.

The separate works did not last long, with sales suffering badly due to an economic slump in 1921. The Galloway name continued but production was transferred early in 1922 to the Heathhall factory, alongside the Arrol-Johnston, where the downturn in sales had left plenty of spare capacity. Dorothée continued to work in sales, though as a result of being in charge of the Galloway stand at the Olympia Motor Show she decided to make a change in career. A newspaper reporter had visited the stand and remarked that she was doing a man out of a job. Despite changes

A brand new Arrol-Johnston Model A tourer outside the factory in 1920. Some of the workers' houses can be seen in the background.

in people's attitudes to women working, brought about by the war, there were still relatively few women in industry, particularly in senior positions. As a result Dorothée moved to Croydon where she used £2,000 of her own savings to start a laundry business which proved to be hugely successful; she felt that washing would not be depriving men of a job.

The 1920s was not the era of golden opportunities which had been predicted at the end of the war and this was particularly true of the car industry. The slump of 1921 was followed by a price war started by Morris Motors which put many companies out of business. Industrial disputes and the high cost of replacing machinery, worn out on munitions production, all contributed to the industry's difficulties. In Scotland at least seven small car firms came and went during the 1920s.

One car which deserved to be successful was the Skeoch, built by the Skeoch Utility Car Co of Dalbeattie, Dumfriesshire. This was a true cyclecar, a design which used a lightweight single-cylinder engine in a light two-seat body, intended as one step up from a motorcycle. Production started in 1921 with the most basic model costing only £165, at a time when the 15.9hp Arrol-Johnston cost £625. Unfortunately barely ten cars had been built before the works was badly damaged by fire. Cost cuttings had included not insuring the works, and production was not restarted.

Perhaps the most successful of the other Scottish makers in the 1920s was Beardmore Motors Limited. Sir William Beardmore had demonstrated an interest in car production with his backing of Arrol-Johnston, of which he was company chairman. After the war he had been looking for a use for a redundant munitions works at Anniesland, Glasgow. A light car was planned using a 15.6hp engine which had been in use as a starter

Partially completed Skeoch cars in the Dalbeattie works which burnt down in 1921.

for the huge Beardmore airship engines. The company was heavily involved in airships and had built the R34 which made the first ever East-West and double crossing of the Atlantic in 1919. In the event, a lighter 12hp engine was specially designed and production of a tourer started in 1922. Several hundred of this expensive quality car were built before the economic downturn saw its demise.

The airship engine planned for the car did, however, find a use in what must have been Scotland's single most successful motor product, the Beardmore taxi-cab. The company had the vision to see the need for a large number of taxis after the war and the design had been completed even before peace was declared. Early in 1919 production started in the old Arrol-Johnston works in Paisley and it became an instant success. The vehicle was specifically aimed at the London market, with advice on the design from Scotland Yard to ensure it met all regulations. By the time that production was transferred to London in 1928, to be near the principal market, over 6,000 had been produced. Various designs of Beardmore taxi continued to be built in London until 1967.

Cars had developed by the mid-1920s into a reliable means of transport, though there were several aspects of design which had still to develop into the definitive shape which would then remain unchanged until the introduction of the Mini in 1959. Perhaps the most noticeable omission was an integral boot, with luggage being carried on a fold-out rack at the rear of cars until well into the 1930s. Wire wheels were still the norm although Arrol-Johnston also used a pressed metal disc style.

At this time British companies had a very poor export record with the entire industry only managing to export about 2,000 cars in 1922. Thomas Pullinger realized the potential of this market and in 1923 made a round-the-world fact finding tour. As a direct result two new models with more powerful engines were introduced by Arrol-Johnston and named the Empire and Dominion to show clearly which market they were aimed at. Neither was particularly successful, though the evidence of examples turning up

Part of the Madelvic factory at Granton, Edinburgh, as seen today, which has changed little since 1899. The carving above the office door depicts a wheel driven by a chain.

The Cragievar Express at Cragievar Castle. The vehicle is now preserved at Grampian Transport Museum, Alford.

A colourful advert for Argyll cars in 1903.

An unusual view of what is possibly the only surviving Arrol-Aster. The car is in the collections of the National Museums of Scotland.

A 1920 Arrol-Johnston tourer from the National Museums collections, on display at Grampian Transport Museum.

The small wheels which caused the Scottish Aviation Scamp to fail on the test track are evident in this photograph of a preserved example.

This photograph of the National Museums of Scotland Hillman Imp recreates an illustration from the 1976 Imp sales brochure.

The Argyll which is built in Lochgilphead, in the old county of Argyll. This illustration comes from the original sales brochure.

Rising star David Coulthard from Twynholm, Dumfrieshire, wearing his distinctive Saltire helmet during the 1995 Brazilian Grand Prix.

today in Australia and New Zealand show that reasonable numbers must have sold. By this time the company was in serious decline despite producing worthy, though expensive, products. The major blow to the company came in December 1925 when Pullinger decided to retire. As managing director he had brought the firm back from near extinction in 1909 and had established the company in a modern factory with modern ideas. He was clearly liked by his employees for whom he introduced many welfare facilities. He retired to Jersey and died in 1945 at the age of 78.

The decline of the industry in Scotland was mirrored in England with the difference that despite a large number of English firms going bankrupt there were plenty of others to carry on; in Scotland by 1925 there were only the three firms in the Beardmore group (Beardmore, Arrol-Johnston and Galloway) and two other independent makes. Of these two, Dunalistair of Glasgow managed to produce only four cars, while the more successful Rob Roy, also from Glasgow, stayed in business long enough to have a reasonable production run. They too went into liquidation by 1926.

Arrol-Johnston were in real difficulties by 1927 and a decision was taken to amalgamate with Aster of Wembley, Middlesex, a long-established engine maker who had entered the luxury car market. Aster was probably seen by the Heathhall management as a way into luxury cars and also as a valuable sales outlet. Sir William Beardmore, by now Lord Invernairn, became chairman of the new company while Mr Lowe and Mr Clench of Asters became joint managing directors.

The first major task of the new company was to rationalize the range of cars which, like Argyll before it, had become complicated. For 1928, the Aster-designed bodywork was fitted to all cars regardless of the marque. Even this was too complicated an arrangement with similar looking Galloway and Arrol-Johnston cars competing for the same market. The rather dowdy image which these names now conjured up was also at odds with the aim of the new management entering the luxury market, and by the end of 1928 they were deleted in favour of the single name of Arrol-Aster.

The range was now cut to only two models, the 17/50 using an in-line six-cylinder engine with sleeve-valves and a larger 23/70 eight-cylinder version. Each was offered with five body styles reflecting the choice expected by the potential customer buying a luxury car in the late 1920s. Despite the new elegant design and a good build quality (though some say that the engine was prone to breakage) the company was doomed. In some respects the management was right to try and enter a niche market; the Scottish industry had always been good at making specialist products. However, the time was wrong for expensive cars with the rise in petrol tax of 1928 and the American stock market crash of 1929.

By the end of 1929 Arrol-Aster had gone into liquidation and although the company remained nominally in business until 1931 under the control of the receiver, only a few more cars were built, probably completing part-built cars and producing spares. The closure of the great Heathhall works effectively ended the indigenous Scottish car industry, even though many more cars were to be built to designs from elsewhere.

After the demise of Arrol-Aster in 1929 there was a brief flurry of activity in Edinburgh where a small company was set up under the name of Scotsman, a name which had been previously used by a Glasgow firm in 1922-23. By now car design was increasingly complex and it would become ever more difficult for small scale production to succeed. To keep costs down a French SARA engine was used, built under licence in Edinburgh. Even the bodies were constructed elsewhere, by the well-known Edinburgh coachbuilders John Croall and Sons.

The new cars were shown at the Scottish Motor Show in November 1929 and were by all accounts well-made sporting models using a 1800cc six-cylinder engine. A smaller 1500cc four-cylinder car was soon introduced, using a proprietary Meadows engine and called the Little Scotsman. Again road tests of the period suggest that these were excellent cars. However, by

An Arrol-Aster competing in the Bournemouth rally. The cars were fast but the delicate engine needed care.

1930 the firm was in financial trouble; these were difficult times for even the large well-established companies such as Austin and Morris. Only 20 or 30 cars had been built when a mysterious fire destroyed the works and nothing more was heard of Scotsman.

In 1931 the last couple of Arrol-Asters were built from remaining parts, and production of the Beardmore taxi in Paisley lasted another year before being transferred to London, where its main market was. There had now been continuous production of Scottish designed and built cars for 33 years. Scottish engineers had been experimenting in motorized transport for a further 115 years before that. The car industry had never been huge, in the way that locomotive and shipbuilding were, but it had provided employment for a considerable number of people. Although probably no more than 25,000 cars had been built in total they

Unsure of whether it was a boat or a car, the Lambert Hydrocar of 1932.

were for the most part a credit to their country of origin with an excellent standard in the quality of build and reliability.

Until the launch of the Hillman Imp in 1963, only a few isolated examples of cars would be built in Scotland. One unusual venture was the Lambert Hydrocar built in 1932 by Leslie Lambert in Bearsden, Glasgow. This was described as 'a novel vehicle' and was designed as a dual purpose car and motor boat. The original sales brochure shows it cruising on the Clyde. It is not known how many were built.

Another obscure make was the Robertson sports car. This venture was started by J W Robertson of Drymen, near Stirling, in 1934. Following in the tradition of Argyll the engines used were of the sleeve-valve design, in this case with a V-4 configuration. The cars were said to have had a top speed of 65mph and a sleek profile. However, after a brief production, and a change of name to Cowal, nothing further was heard of the make.

There are many reasons behind the absence of a car industry in Scotland during a period which saw car production elsewhere expand rapidly. Large companies such as Austin in England had become well established in mass production by the time the Scottish industry had failed. To restart a new industry, in the face of such opposition, would have been an uphill struggle to say the least. Another problem which had always beset the Scottish firms was that the support industry, producing carburettors, electrics and a host of other parts, was centred on the English midlands. This made supply more difficult, and the parts more expensive because of transport.

Production methods were also changing as cars became more complex. This involved a much greater investment in new tooling and a change to mechanized production lines. During the 1930s the shape of car bodies became more rounded, and the flat rear end was replaced with the complex curves of an integral luggage boot. These body panels required larger and more powerful presses to form the shape from sheet steel. The huge investment required would gradually reduce the number of car makers world wide.

With the outbreak of World War II in 1939 all production of private cars halted and car factories were once again turned over to the manufacture of munitions and aircraft. In Scotland there were no car factories to convert, but Albion were still successfully building commercial vehicles and turned to making military trucks and handguns. The excellence of Scottish engineering was not forgotten and new factories were built to utilize that expertise. The new Rolls Royce aircraft engine factory at Hillington near Glasgow was the largest engineering works in Europe during the war. After the war Albion continued successfully building commercial vehicles, but there was no sign of any attempts to revive car production. However, this was due to change dramatically as government plans to relieve economic blackspots throughout Great Britain focused on the west of Scotland.

6 Linwood, a new factory

Follow the Suez crisis in 1956, when petrol became scarce and expensive, there was a sudden rush by most of the major car makers in Europe to produce designs for new small and economic cars. The market leader was undoubtedly the Austin/Morris Mini launched in 1959. Alex Isigonis, the Mini's creator, was once heard to say that his new car was going to remove the curse of the bubble-car from our roads. Indeed it was the motorcycle and sidecar, along with the Italian bubble-cars, which had made the country mobile again after the end of the war, and it was this market at which cars like the Mini were aimed.

One company which aimed to take a share of this new market was Hillman which, along with Humber, was part of the Coventry-based Rootes Group. The group had been formed in 1932, just two years after the demise of Scotland's own industry, by William and Reginald Rootes who were in the retail car business running a chain of showrooms throughout England. The brothers had become shareholders in Hillman and Humber with the aim of gaining some control over the cars which they were selling, but when the two companies faced bankruptcy they took full control. After reorganizing and introducing an entirely new model, the Hillman Minx, they revived the fortunes of both marques to become a major force in British car making.

Further acquisitions were made including Sunbeam and Clement-Talbot. Much of their success was due to the use of common components in several makes of car. Different marques were combined as it suited them; Sunbeam-Talbot was formed to market a more sporty line of cars. The staple output of the group was, however, upper-range family cars, the most notable of which was the worthy Hillman Minx. After World War II the one type which was still absent from the Rootes catalogues was a small car to compete with the likes of the Morris Minor or Austin A30. It was with this in mind that two young engineers working for Humber sought permission to develop a new light car.

Tim Fry and Mike Parkes were enthusiastic about producing a rival to the bubble-car more in the line of a 'proper' car able to carry a small family. Work started in 1955 and, despite the end to wartime petrol rationing, and before the Suez Crisis, they aimed at producing a car capable of at least 60 miles per gallon. A prototype, known as the Slug, due to it's ugly appearance, was duly built and presented to management for approval. This initial design was turned down but the project continued, known thereafter as the Apex, and with even more commitment from Fry and Parkes if not the company.

The first design had incorporated a small two-stroke engine but the general noise and lack of sophistication ruled it out and the car was developed around the more conventional notion of a four-cylinder four-stroke engine, though in most other respects the power plant was quite different from anything in use at that time. The chosen design came from Coventry Climax, a company who had an excellent reputation for building lightweight fire-pumps. They were developing a new all-alloy engine and had built a version for possible use in cars. Some of these units had been successfully raced and it was a V8 version of this engine which went on to become famous for winning Formula I races.

With a power plant chosen, it was now possible to finalize the design in the form which was soon to be known as the Hillman Imp. Many features of the car were unusual; not only was the engine all-alloy but it was installed inclined at a 45 degree angle, in line at the rear of the car, when the norm was to find it in the front, upright and made of steel. Design work continued through the late 1950s, and was still ongoing when the Mini was launched in 1959. At this time it was decided to put the Imp into full production and investigations were made into whether there was enough capacity to build it at the main Rootes plant at Ryton on Dunsmore near Coventry.

It was evident that more space was needed and the first choice was to extend the existing factory on adjacent land which was available. The government had to agree to this scheme and it

The new Linwood factory linked to the Pressed Steel plant beyond by a bridge over the public road. The town of Paisley appears in the top right of the photograph.

soon became clear that they had other ideas. A policy for regional development was being implemented with the aim of creating employment in blackspots such as Liverpool and the West of Scotland. Government money was available to encourage dispersal of industry, and anyhow Rootes were flatly refused permission to expand at Coventry.

Lord Rootes entered into informal discussions with Prime Minister Macmillan about various possible locations in the North of England and Scotland, and it soon became clear that a site in Scotland was favoured. This was partly influenced by the government having invested heavily in the steel plant at Ravenscraig in Lanarkshire which was still working well below capacity. The decision on a specific site was no doubt made on the basis of the location of the Pressed Steel plant already established to produce car bodies on a site at Linwood, some fourteen miles west of Glasgow. This factory had originally been built just after the war by Beardmore. A government loan was arranged at preferential rates and construction of the new Linwood car plant was started in April 1961, the contract having been given to a local Glasgow firm.

The new factory was everything that a modern car plant should be and was officially opened by the Duke of Edinburgh in May 1963. The new building was separated from the Pressed Steel plant by the main Linwood to Paisley road, but even this minor difficulty was catered for with the provision of a bridge carrying an overhead conveyer between the two factories. A rail link connected with a branch from the main line at Ferguslie Park, straight into the factory complex, reflecting what Argyll and Arrol-Johnston had done when they built new works. In this case the rail link was essential because production would far exceed that of the entire Scottish industry from before 1930, and there were some curious working practices to be catered for.

The half-built steel shell of an Imp on the production line in the Pressed Steel factory.

The strangest of these arrangements came from the way that the engines were to be produced. A new state-of-the art die-casting facility had been incorporated in the Linwood plant to manufacture aluminium alloy components for what was to become the first all-alloy engine in a British mass-produced car. These alloy parts were then sent to the main Rootes factory at Coventry where the steel components were made, and the complete unit built up together with the cylinder head which came from Birmingham. The finished engines were then sent back to Scotland to be installed in cars which, in many cases, were sent back for sale in England.

The day after the official opening of the factory the first Imps went on sale to the public, little more than two years after construction of the factory commenced. This incredible rush to put the car into production was ultimately to contribute to its demise some thirteen years later. Although there were indeed cars ready for sale on 3 May, there were still many problems which could have been rectified had the launch been delayed and more time for testing allowed. As a result the early cars gained a reputation for unreliability which was to taint the Imp throughout its production life.

Much of the testing was undertaken on roads in the Scottish Highlands, in many cases the same roads which had witnessed the very first public reliability trials 60 years earlier. Although the tortuous Highland roads made an excellent proving ground there were many aspects which could not be tested here. A determined export campaign was planned and as much of this market would be Europe several cars were sent to be tested at high altitude in the Alps. These tests were designed to show up any deficiencies in the cooling system, brakes and carburation.

Perhaps due to the short time allowed, or perhaps just bad luck, very few of the problems which gave the Imp a poor reputation actually showed up until the cars went on sale to the public. Even the first test reports appearing in the motoring press gave no hint of the problems ahead, with most writers finding that it exceeded their expectations and few finding any notable faults.

The Scottish car magazine *Motor World* even penned a poem celebrating the launch of the car which they saw as the beginning of a new era in Scottish car building - even if it was an English design built by an English company.

Despite the English origins of the car it was soon seen by the people of Scotland as their car. Indeed, by February 1964 when the new Prime Minister, Sir Alex Douglas-Home, visited Linwood over 76% of the car was being produced in Scotland and importantly by a predominantly Scottish workforce. There was certainly no shortage of unemployed Scots desperate to get a job, any job. It was here with the workforce that many of the later problems had their roots. The initial group of engineers who commissioned the factory had been sent North from the parent firm in Coventry and they were expected to source the workforce locally, with additional experienced engineers being brought up from the Midlands as necessary.

The old meets the new. On the left is Mr Alex Wise, Imp test driver, who worked for Arrol-Johnston, being introduced by a Rootes sales executive to Mr Andrew McBride who worked for Argyll.

Very few experienced men were willing to be transferred to Scotland, and there was almost nobody locally with a background in modern production engineering. Even among the workers taken on for work on the shop floor, few had worked in any kind of factory before, with by far the largest group being ex-shipyard workers. The methods used in building a modern motor car were quite alien to these people, even though they were highly skilled in heavy engineering. The working practices in the shipyards were also quite different and were to be the basis of the industrial relations problems later on.

However, in May 1963 these problems were still a long way in the future and the royal visit to open the factory, along with the excellent press reports, gave the impression that the new car could only be a complete success. In fact there were only about 800 completed cars available by the launch, which was far too few to satisfy the public demands for a car which had received such good publicity. Although the Imp cost considerably more than the Mini in 1963, £508 compared to £448, it was more spacious and had greater sophistication which would appeal to a wider public. It should have been serious competition to the Mini, but British Leyland already had four years of sales by the time the Imp was launched and the Mini had few faults.

Of greater concern to the workforce in 1963 than problems with production was the problem of where to live. The small town of Linwood had a population of less than 15,000 prior to the advent of the Rootes factory. Although the larger towns of Paisley to the east and Johnstone to the west were almost adjacent to the factory there was little room for expansion. Linwood was on the other hand surrounded by open land and was still almost on the factory doorstep. The local council implemented a crash programme of house building as the population grew by over 7,000

One of the many schemes devised by the Rootes management to encourage a happy workforce.

ROOTES EMPLOYEES'

CONTINENTAL MOTORING HOLIDAY

'SWITZERLAND'

From Monday, July 18th to Sunday, July 31st 1966

THIS COULD BE **YOU** IN YOUR NEW CAR ENJOYING YOUR OWN SPECIAL CONTINENTAL MOTORING HOLIDAY

Try out your car on the traffic-free and scenic Continental roads and give your family a fabulous Swiss holiday on the shores of the picturesque LAKE LUCERNE.

FANTASTIC VALUE FOR MONEY — SPECIAL TERMS APPLICABLE TO THIS UNIQUE HOLIDAY — ONLY FIRST CLASS ARRANGEMENTS AT MODERATE COST.

SPECIAL ATTRACTIONS include a gay night in BRUSSELS, a visit to the CHAMPAGNE CELLARS and many other excursions and get-togethers. A linguist TOUR MANAGER to take care of all your problems, if any, and to arrange your entertainments – but NO REGIMENTATION, CONVOY DRIVING or 'COACH-PARTY' TREATMENT!

READ THE ITINERARY, SEE ALL THE FACILITIES PROVIDED AND THEN COMPLETE YOUR BOOKING FORM.

ALL FORMALITIES REMOVED, THERE IS NOTHING MORE FOR YOU TO DO, BUT ENJOY THE HOLIDAY!

A large number of women were employed at Linwood; in this shot they are finishing gear casings.

in only four years. The infrastructure of the area also grew rapidly with new shops and facilities, though to begin with this growth could not keep up with the population.

Many of the workers came from Clydebank, just across the River Clyde, though even this proved to be too far for people to commute, with most relying on public transport and even those few who had a car finding the lack of crossing points on the river to be a hurdle. The new workforce also consisted of a large number of women so it was only natural that whole families should move to the area. The Linwood management were only too keen to foster a good relationship with the new workforce and many welfare schemes were introduced. A Family Day at the factory was instituted to allow workers to bring friends and relations, especially children, along to see where they worked and how the cars were built. Staff facilities were excellent with a canteen as good as that of any other factory in Scotland.

7 Linwood in production

For the first few months things went well for the Imp, with over 30,000 cars being produced by the end of the year and only eight strikes halting production, but by 1964 things were starting to go wrong. The reliability problems had now become apparent, with a catalogue of faults particularly with the new all-alloy engine, which although of sound design was relatively untested, like the rest of the car. Overheating became one of the worst problems, leading to blown cylinder-head gaskets and distortion of the main engine casting. These were major problems causing breakdowns, and often led to major parts of the engine requiring replacement. The body of the car also suffered problems, with broken transaxles and the steering kingpins, supposedly sealed for life, seizing and causing stiff steering or even the dangerous collapse of the front end of the car.

The final assembly track where Imps were fitted with head-lamps and other small parts.

Not everything was doom-and-gloom with the Imp however, with export orders beginning to look encouraging and some useful publicity gained which helped to cover up the increasing number of poor reports. Early in 1964 the new Grand St Bernard tunnel in Switzerland was opened and a Hillman Imp was chosen to be the first car officially to drive through. The British School of Motoring placed an order for 50 Imps which, as well as being a welcome bulk order, was an excellent way of publicizing the car, always provided that there were not too many breakdowns.

From the beginning the Imp was available in two versions, the Standard and the Deluxe, with the Standard being replaced by an improved specification Super in 1965. Although these models were to be the staple production, from 1963 until the end in 1976, there were other versions which included a coupe and a van. The first variant to be introduced was the Sport, intended as the car to attract the younger driver. Unlike the Mini, the Imp had always been seen as a whole range of cars rather than a single model. Rootes were also keen to exploit their, by now common, use of so-called 'badge engineering' where different marques denoted either a luxury or sporting image. The Sunbeam name was applied to the Sport Imp which for 1967 became the Sunbeam Stiletto, a sport coupe with four headlamps and special wheeltrims.

An estate version known as the Husky retained the Hillman family name but the van version took the Commer name to fit in with the range of larger commercial vehicles produced by Rootes. Both the van and the estate shared one unusual feature which added to the quirkiness, and which perhaps had not been considered originally, which was the access to the engine. Unlike the car these versions had a lifting panel in the load area, necessitating an empty vehicle if the engine was to be worked on. Allowance was made for servicing the oil and water without lifting the lid with access holes at the rear.

By the end of 1964 it was obvious that it had been a bad year at Linwood. The eight strike stoppages of 1963 had risen to 31 and production had only reached a total of 50,000 for the year; far short of the capacity to build 150,000 units per anum. Both Rootes and Pressed Steel were forced to go on to four-day working and make redundancies to help the company finances. The Scottish factory continued to be a drain on resources and late in 1964 they were forced to sell shares to the American car giant Chrysler. Two years later the Americans took a majority stake in the company and began to use their influence to make cost-saving changes resulting in a further decline in build quality which did nothing for the Imp's reputation.

Some Imps had been exported to the USA, with the very first going to film-star Cary Grant, but it was not well received there, being seen as a funny little European car. Neither Chrysler management nor the American public understood the Imp, and production only continued because there were no plans for another small car to replace it. Production steadily declined from a high point in 1964, to only 19,000 cars built in 1970 and a mere 10,000 in 1975, the last full year of production.

There were profitable years at Linwood but these were more than balanced out by the years when there were large losses. In 1973 for example the plant was in profit by over £4 million only to return to a £5.8 million loss the following year. In 1975 Linwood contributed an £8 million loss to the £35 million loss made by the parent company. These losses could not continue and the last Imp rolled off the end of the production line in March 1976. A total of 440,032 Imps had been built, which was barely 25% of the potential two million car output.

This was not, however, the end of car production at Linwood. As part of their major shakeup of the Rootes Group, Chrysler had transferred production of the Hillman Hunter to Scotland and this had been built side-by-side with the Imp. Now the additional capacity on the production lines was utilized by moving the Hillman Avenger assembly line north from Coventry in 1976. This move kept the Linwood factory alive but was seen by the Scottish workforce as a snub. The new Sunbeam Alpine was to be built at the Coventry factory and the Avenger, now an outdated model, had to move to make way and was dumped on Linwood.

One last new model was now planned for manufacture at Linwood, the Chrysler Sunbeam. Unfortunately this was not a new state-of-the-art car which would take the market by storm. The Sunbeam was in fact cobbled together using Avenger running gear with a new body shape. Even the engine was simply an enlarged version of the Imp unit which at the maximum size of 930cc was too small for the larger car. The public were not fooled and sales were poor.

One final twist in the tale for the Linwood plant occurred at the end of 1978. Chrysler had by now been suffering crippling losses from its UK subsidiary Rootes and a French subsidiary Simca. As a result both companies were sold to Peugeot-Citroen for £225 million. What had been Chrysler Europe now became Talbot, taking the name from the Clement Talbot company bought by Rootes back in 1935. This now left Linwood, including

Not everyone's idea of a Scottish car but the Avenger was built at Linwood from 1976.

the Pressed Steel Plant which had been acquired by Rootes some years earlier, in French hands.

Most workers at Linwood knew that this was the end. With two of the Scottish-built cars nearing the end of their production life, and the third with poor sales it was obvious that the French had no interest in a factory remote from any other of their interests. The final blow came when the French Government offered Peugeot-Citroen two brand new factories, free.

On 22 May 1981 the entire Linwood complex, including Pressed Steel, closed making nearly 5,000 workers redundant. What had started as a British Government move to help employment in a depressed area, now ended with the creation of a new unemployment blackspot in the West of Scotland. Linwood was further insulted when, the day after closure was announced, the government revealed that it was to give the American John DeLorean a further £10m for his car factory in Northern Ireland, bringing the total invested to £80m.

Today the remains of Linwood make a depressing sight. The Pressed Steel factory has been partly demolished and is used as a retail park. Much of the Rootes factory is still standing and now subdivided into small industrial units. The land around the plant, intended by the government for ancillary industries, still lies empty.

8 The modern era

While the Hillman Imp and the Linwood factory were making the headlines throughout the 1960s and 1970s there were other small-scale projects underway elsewhere in Scotland. None would ultimately lead to production-line quantities but each was an important part of the story of Scottish cars nonetheless.

The first new venture after the opening of Linwood in 1963 came unexpectedly from the aviation industry. Scottish Aviation at Prestwick, on the Ayrshire coast, had spent the war years repairing and modifying thousands of aircraft and as a result had built up a massive workforce with engineering skills. After the end of the war the volume of work declined rapidly. Despite building their own design of aircraft the company had to diversify to stay alive. It first became involved with road vehicles by building aluminium bodies for vans and trucks. This led to the Scottish Aviation Project Department becoming involved in the design of a car; the first to be conceived in Scotland since the Robertson of 1934. The new car was a return to an old idea, that of electric power.

Work started late in 1964 and by early 1965 a test vehicle nicknamed 'the farm cart' was running successfully. This showed real promise, with a top speed of 36mph and a standing start to 30mph in only ten seconds. One of the biggest problems with electric power had always been limited range, but even this looked good on the test vehicle. Urban driving gave a range of 18 miles before re-charging.

The aim of the team was to produce a car for urban commuting which would be cheap to run and nearly silent. Most journeys to work were of less than ten miles and made with only one or two people in the car. This determined the specification calling for an enclosed two-seater with good manoeuvrability and a range of about 30 miles. Overnight charging would give the benefit of cheap off-peak electricity.

The car was called the Scamp, a name which combined its Scottish origin and electric power. Work had progressed to a point where in July 1965 Scottish Aviation entered into negotiations with the Electricity Supply Board to market the new car through their 2000 showrooms. The board were impressed with the initial results and the next step was to construct a body using wood and aluminium, and register the car for use on the road.

The prototype vehicle was taken to London and Bristol on a series of demonstrations which prompted a large number of enquiries from potential customers at home and abroad. The car was well received and publicity was good, particularly when the racing driver Stirling Moss drove one. A further twelve cars were now built for testing, the first of these being shown at the Ideal Home Exhibition in February 1967. Everything looked set for the company to start series production of this alternative to the conventional internal-combustion car.

Unfortunately, as with the earlier Scottish car industry, things began to go wrong for the project. The first problem was with the batteries available at the time. Lucas had provided sets which were able to give the required range between charges, but they were only lasting twelve months before needing to be replaced. It was felt that the minimum economic life should be eighteen months, and although Lucas thought they could achieve this the cost would be prohibitive.

The final factor which probably caused the project to be cancelled came when the Scamp was submitted to the Motor Industry Research Association for evaluation. The tests which were applied to the Scamp were the same as those given to a conventional car. Clearly this small two-seat commuter vehicle using tiny eight-inch wheels was no match. Eventually when the suspension broke on a bumpy test track, the car was claimed to be unroadworthy. Although improvements could have been made, this was the final straw and Scottish Aviation abandoned the Scamp to return to aviation work.

Clearly it was unlikely that any further volume production of cars would return to Scotland. Even the mighty Rootes Group and Chrysler had failed to maintain production despite having a new factory using the very latest computer-controlled methods. There was, however, an opening for small specialist builders to establish themselves in Scotland with the help of government grants.

The first new specialist firm in Scotland was established to take over manufacture of the failed Probe car from England. The Probe had been designed by Dennis Adams, stylist of the Marcos sports car, as an exercise to build the lowest possible car. When it was launched in 1969 the 29 inch (74cm) height of the Probe 15 was greeted with amazement. It was so low that access was through the roof. Interestingly, the car was powered by the Hillman Imp engine which was ideal because lying at a 45 degree angle it fitted into the low profile. This was only one of many specialist cars to use the Imp engine. A single example of the Probe 15 was built, with a further three of the more practical Probe 16. A new version, the 2001, was produced, but the company went into liquidation shortly after. At this stage the Probe was bought by a Scottish firm based in Irvine, on the Ayrshire coast. Here twelve cars were produced before this company also failed in 1972.

One of the more unusual cars to be produced in Scotland was the Parabug. This was built in Aberdeen for a short time around 1972. Designed as a semi off-road car it was based on Volkswagen running gear and used a simple open glassfibre body with soft top. The design did not last long against stiff opposition in this limited market.

The next attempt at building specialist cars in Scotland came in 1984 when AC (Scotland) was formed at Hillington, near Glasgow, to build a mid-engined sports car. This project had been purchased from the well known firm AC, of Thames Ditton, Surrey after they had taken five years to develop it. The car was renamed the AC Ecosse, but the now familiar story of failure was repeated when in October 1985 the receiver was called in.

The Aberdeen built Parabug off-road car, in an environment for which it was designed.

Once again the company was reformed, this time as Ecosse Cars. Again failure followed and the company moved south to establish production at Knebworth, Hertfordshire. The failure of the Ecosse as a Scottish product was particularly unfortunate because of the potential this car had to compete with Porsche and Lotus. The company had signed a deal with Alfa-Romeo for the supply of engines and this would have opened up an all-important network of dealers to market and service the car. The project deserved more success.

Perhaps the most significant Scottish car of recent years has been a high-performance sports car reusing the old Argyll name. This company has perhaps more right to this famous name than their predecessors, being based in Lochgilphead in the old county of Argyll. The company was started by Bob Henderson who already ran a successful business called Minnow Fish producing high quality carburettors. Based on a Canadian product these improved versions are said to be capable of outliving the engines to which they are fitted while also improving performance. Henderson's company became the first in Europe to manufacture turbo-chargers for cars.

With this background in components for high performance cars, the move into building his own supercars was a natural progression for Henderson. He completed the first prototype Argyll Turbo in 1976 but then took another six years to put the car into series production. Development costs amounted to only £250,000, a small cost for a car which would be capable of competing with Ferrari. Unlike almost every other recent car project in Scotland, particularly Rootes, this was achieved without government financial help.

The car was finally launched in 1983 by the Duke of Argyll in the grounds of Inveraray Castle.

The design of the car was pure Scottish engineering excellence which lived up to the reputation of the original Argyll as a quality product. The chassis uses a combination of box sections and a tubular space frame giving a rigid structure similar to many racing cars. The first production cars used a 2.7 litre V6 Peugeot engine which gave a top speed in the region of 140mph. This was later replaced with a 4.2 litre V8 engine offering a possible top speed of 180mph.

Even the styling of the body is entirely Scottish although some parts, such as the Volvo dashboard, are bought in. The body of the car provides an interesting connection with the early Scottish car industry. The mould, or plug, used to produce the glassfibre body was made by a company called Solway Marine based in a building next door to the old Arrol-Johnston works at Heathhall.

The most noticeable feature of the car is the length, a result of the need to have two seats in the rear as well as the engine mounted in the middle of the car. Most cars in this class can barely accom-

The Haldane HD 100 which was built in Blantyre and East Kilbride bore more than a passing resemblance to the Austin Healey 3000. An 1980s car with 1950s styling.

modate one child in the back. The long wheelbase does ensure a good ride and the mid-engine gives safe neutral handling.

Although the Argyll Turbo is an expensive car, even in its class, it is designed to last. The glassfibre body will never rust and the hand-finished interior should ensure a longevity which can only be dreamed of with today's mass-produced cars. The Argyll has a niche market and will never rival even the original Argyll for numbers produced. However, a steady sale of cars to the United States makes this Scottish product an export success.

The story does not quite finish here. In 1995 an Aberdeen company has built the first of a new marque called Ascari. The founder is racing driver Klaas Zwart and the car has been designed by Lee Noble. This supercar will challenge the best that the Italian industry can produce. A racing version of the Ascari has had its first outing and looks set be a winner. A decision has not been taken as to where the cars will be built, but it is hoped that the Ascari will join the Argyll to continue the 100-year-old tradition of cars made in Scotland.

9 Relics of an industry

Gone are the hordes of workers streaming out of Linwood, the bustle and noise at Alexandria and the whir of machinery at Heathhall. But today, despite the passing of the years, there is still much evidence to show that Scotland once had a motor industry. Many of the factory buildings have survived to find other uses. Even the cars themselves have survived in reasonable numbers.

Some of the factories retain a connection with the motor industry, albeit tenuous in most cases. The closest that any of the buildings come to car production today is that of the Albion works at Scotstoun, on the banks of the River Clyde. Here even the name lives on, with Albion Automotive producing axles and other components for commercial vehicles. After a period as part of the former Leyland DAF Group it is once again an independent Scottish firm.

Also to be found in Glasgow is the steel-framed Truscon building built in 1913 for the firm of William Park, a well-known coachbuilder of the time. Of all the buildings related to the motor industry in Scotland this one must be unique as it is still used for virtually the same purpose, as the coachbuilding and accident repair shop of a large car dealership.

Some of the factories can still be seen in the same condition as when the last car rolled out of the despatch bay doors. The best example is the Arrol-Johnston works at Heathhall, on the main Dumfries to Moffat road. For many years the factory was owned by the North British Rubber Company producing the famous Hunter green wellington boot. Now used by the Gates Rubber Company Ltd, there is little to show the passage of time. A few buildings have been added, and some of the company's workers' houses demolished. On the opposite side of the road from the main building can be seen the former canteen block, also built as part of Pullinger's welfare facilities for the workers.

While the main Arrol-Johnston factory survives in perfect condition, the subsidiary Galloway works has seen considerable change. Although the picturesque rural surroundings have changed little over the years, the windows of the daylight factory have now been replaced with brick infill giving a derelict look to the site. Now used for battery egg production, there is nothing to show that motor cars and engines for fighter aircraft were once built here. The weigh-house still stands at the gates through which the Galloway cars passed, but close inspection shows that the weigh-bridge is a later replacement.

Possibly the most famous of all the factories, and certainly the grandest, was Argyll's palatial building at Alexandria near Loch Lomond. When driving past the site today the grand facade stands witness to the financial folly of a great company. The building lies empty, awaiting a new use, while the main factory building behind has now been demolished. The only evidence that cars were ever built here can be seen in the carving above the main entrance where classical figures surround an Argyll car.

In Edinburgh, fortune has been kinder to the Madelvic works, the first purpose-built car factory in Britain. Like the Argyll building, evidence of its origins can be seen with the carving above the office door showing a chain-driven wheel. Although now considerably extended, the original 1899 factory survives almost intact, though the circular test track has been lost under the later developments. The site has been owned since 1925 by the United Wire Company who have made good use of William Peck's original £33 000 investment.

The site of Scotland's largest car plant, built on the Linwood boglands, is today largely wasteland once again. The Pressed Steel plant has almost vanished, with piles of concrete rubble marking the site. The main Rootes plant survives in parts, though most of the buildings have been subdivided into industrial units.

And of the vehicles themselves, it is now 65 years since the last wholly Scottish car rolled off the end of a production line, yet the cars survive in surprisingly large numbers. Over 120 are known to exist worldwide, and many more may lie undiscovered in barns and garages. There are Albions in South Africa, Arrol-Johnstons in New Zealand and Galloways in Australia. The Argyll must have been popular in Ireland as several examples remain there. Of the four surviving Stirling cars, built in Hamilton, two can be found in England.

There also remains a solitary example of the ten Dalgleish-Gullane cars built in East Lothian around 1908, a better survival rate than that of most of the small companies. Another single survivor is the Arrol-Aster 17/50 to be found in the collections of the National Museums of Scotland. Most vintage cars have interesting stories behind their continuing existence. In this case the car had been used as a taxi and following an engine failure had been pushed to the back of the garage. It was then used as a changing room for the mechanics before being rescued by an enthusiast.

The Albion marque accounts for about 25 of the extant Scottish cars. Argyll is well represented with over 30 survivors, almost all from the great years of the firm between 1899 and 1914.

Only one example of the Argyll cars built after World War I has survived. By far the most numerous of all the makes is Arrol-Johnston, which is perhaps no surprise since they probably built more than any other company. What is surprising is the fact that nearly 15 of the Arrol-Johnstons to be found today are examples of the magnificent wooden Dogcarts. Many of these striking vehicles are still driven on the road, a tribute to the quality of workmanship and materials used. In fact, fine examples of most models of Arrol-Johnston are still maintained in roadworthy condition. One example can still be found just a few miles from the Heathhall factory. This car, a 15.9hp with an All Weather Tourer body, is owned by the Royal Scottish Automobile Club and kept in running order by Ian Gray of Dumfries.

Glasgow was the birthplace of the three largest car firms and it is therefore appropriate that the Museum of Transport in

One of several Albions in the National Museums of Scotland is this 1907 wagonette, seen here in 1938 at the Empire Exhibition in Glasgow.

Glasgow should now be home to the largest collection of Scottish cars to be found anywhere. There are examples on display illustrating most periods of the Scottish industry including the Hillman Imp driven by the Duke of Edinburgh at the launch of the car. This fine display is a must for anyone with an interest in Scotland's past.

Scottish cars can also be found in other museums in Scotland. After Glasgow's Museum of Transport, the next largest collection is that of the National Museums of Scotland in Edinburgh. Grampian Transport Museum at Alford and Myreton Motor Museum at Aberlady usually have examples on display. The small town of Biggar, some 25 miles from Edinburgh, was the birthplace of John Blackwood Murray. Here the museum holds the archives of the original Albion company, a dogcart in running order, as well as the lathe used to make the parts for the first car. They also preserve a few of the large number of Albion commercial vehicles which survive. The most common place to see a Scottish-built car today is still on the public roads. Many Hillman Imps are in daily use and an active owners' club will ensure that these little cars will be seen on the road for many years to come.

10 Motorsport

The Scottish people have always been competitive by nature and this is typified in the field of motorsport. Few competition cars have been built in Scotland and even fewer have ever aspired to international success. However, Scottish-born drivers and teams have more than made up for the paucity of cars, by regularly taking on, and beating, the rest of the world. The cars that have been built deserve some recognition. Probably the first racing cars to be constructed in Scotland were the work of the large engineering firm G & J Weir Limited of Cathcart, Glasgow. William Weir had been a keen motorist from the pioneering days and when he was offered the chance to build three cars to compete in the 1904 Gordon Bennet Race he grasped the opportunity.

The French company Darracq were keen to expand their market and decided to enter three cars in the race as British vehicles. To qualify, the French-designed cars had to be actually constructed in this country and, after some difficulty in finding a builder, Weir were contracted for the job with only ten weeks left. The work was supervised by none other than Thomas Pullinger who at that time worked for Darracq, long before he moved to Arrol-Johnston. The three Weir-Darracq racers were completed on time to be weighed and given approval by the Royal Automobile Club in London. Although the three cars failed in the actual race the work was not completely wasted as the cars went on to achieve success in other events.

In contrast to the failure of the French-Scottish cars, the first all-Scottish effort was rewarded with success at first attempt. John Napier of Arrol-Johnston entered two of his company's cars in the 1905 TT races in the Isle of Man, two years before the first motorcycle TT races. Napier himself was the winner in one car, narrowly beating a Rolls-Royce, while the other car came in fourth. Ironically,

John Napier winning the 1905 TT races on the Isle of Man, driving an Arrol-Johnston.

amongst the cars which Napier beat were the Beeston-Humbers entered by Pullinger who at the time worked for that firm.

This was the first and only win by an Arrol-Johnston. In 1906, and again in 1907, both cars failed to finish the event. There were attempts to succeed in Grand Prix racing, with three cars entered in the 1911 French event resulting in two failures and a ninth place overall. In 1912 the cars were again entered for the French Grand Prix though this time the cars failed to even start the race. This was one failure too many and the company withdrew from racing for good.

Scottish cars had more success at record breaking than at racing. In May 1913 a specially built Argyll was sent to Brooklands race track in the south of England to attempt the twelve hour record. On the day the car ran so well that the run was extended to fourteen hours and resulted in the capture of 26 British records. The car had maintained an average speed of over 70mph. A week later the same car was used to break a number of world records running at an average of over 76mph.

The specially built Argyll car which captured 26 British records at Brooklands in May 1913.

Probably the greatest success for a Scottish constructed car, in speed records, came in 1928 with the Napier Arrol-Aster. This car had been commissioned by Sir Malcolm Campbell for an attempt on the Land Speed Record. The car was built at the Heathhall factory in 1927 and shipped out to Daytona in the USA during February 1928. The car set a new record at 206.96mph (333.06kph) only to be beaten by the Triplex Special two months later.

All the major companies had been involved in motorsport in one way or another, from record breaking to hill-climbs and TT racing to rallies. Apart from the occasional commission none of the cars were produced for sale and were purely used in promoting the makers names as builders of reliable cars. It was not until 1950 that the first Scottish company was established solely for the

In 1928 Malcolm Campbell made a new Land Speed Record driving this 'Bluebird' with the body built at the Heathhall works of Arrol-Johnston & Aster Engineering.

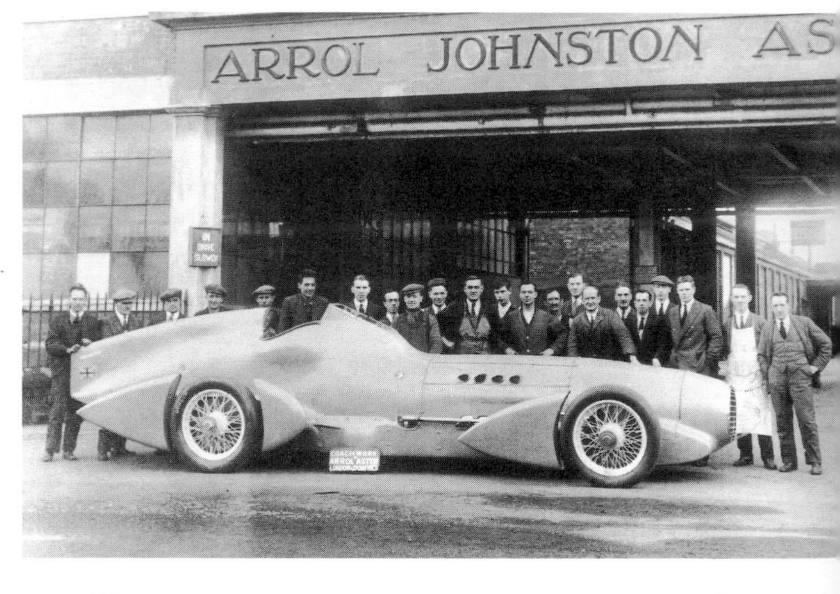

manufacture of racing cars. This was started by Joe Potts, a well-known engine tuner, at Bellshill in Lanarkshire. The cars were small single-seaters designed for 500cc Formula 3 racing and utilized single-cylinder motorcycle engines in a tubular space-frame chassis. Some 30 to 40 examples were built between 1950 and 1954. Previous to this there had been occasional specials built by individuals. James Anderson, a garage owner from Newton Mearns, built four competition cars over a period of twenty years. He had a number of successes in off-road trials during the 1930s, and the last car that he built, in 1936, still survives in the Museum of Transport in Glasgow.

Perhaps the most consistent motorsport success story of any car built in Scotland was that of the Hillman Imp. Not only did private owners build specials to race, but Rootes dealers such as Hartwell produced modified Imps for sale as race cars. However, it was the success of the early works-entered cars which gave the Imp its greatest wins. After a disappointing start in the 1964 Monte Carlo Rally the factory entered again in 1965 and gained valuable publicity when Rosemary Smith from Dublin came 22nd overall and fourth in the 1 litre class, a good result in a tough rally renowned for retirements. 1965 was to be a good year for the Imp starting with Colin Malkin winning the Circuit of Ireland and Rosemary Smith coming second. The following week Rosemary won the Tulip Rally with another Imp second. In the tough RAC Rally, Rootes won the team prize. There was major success on the race-track also. Bill McGovern won the prestigious British Touring Car Championship three times in a row in 1970, 71 and 72. Even today, nearly 20 years after the last Imp was built, the cars are still winning, though now in historic car races.

The geography of Scotland has probably contributed in part to the popularity of hillclimb races, rather than those held on circuits. Before World War II there was no equivalent North of the Border to the famous Brooklands track in England. The hill-climbs had their origins in the reliability trials held in the early years of the century. Each trial included a number of famous hills where success lay in the ability to reach the top of the hill, no matter how slowly.

As cars improved it became more a matter of speed and some of the great hillclimb events were born. Many were in remote areas such as Cairn o' Mount in Kinkardineshire and Rest-and-be-Thankful in Argyll, but huge crowds found their way there to spectate regardless. The Rest-and-be-Thankful run continued to use the original road until the 1970s, long after it had been bypassed. Not all the venues were remote, with the Bo'ness Hillclimb taking place in the grounds of Kinneil House on the edge of this West Lothian town. After World War II there were at least ten hillclimb venues at various times, some being used for the British National Championship. A popular event today takes place on Lord Doune's estate at Doune near Stirling.

To make up for the lack of tarmac circuits in Scotland an alternative was found by using coastal sands at low tide. One of the

Racing on the sands at St Andrews in the 1950s.

most popular locations between the wars was the sands at St Andrews in Fife. The shallow beach was ideal, with the water receding a long way and leaving firm sand. It is difficult to imagine any of today's racing cars being used on a beach.

During World War II the crash programme of military airfield construction saw over 30 new bases in Scotland and many of the pre-war grass strips had concrete runways laid. At the end of the war most of these airfields closed just as rapidly as they had appeared, leaving a legacy of derelict buildings, runways and perimeter taxiways. It soon became apparent that the twists and turns of the miles of taxiways would make ideal instant race tracks. The rural location of most of the airfields would ensure there was little noise disturbance and the acres of grass would be ideal for spectators. During the 1940s new tracks were opened up, from Crimond in Aberdeenshire to Winfield near Kelso, just one mile from the English border.

Scotland's first tarmac racing circuits proved to be immensely popular. The second meeting to be held at Winfield drew a crowd of over 40,000. This was all the more surprising given the continued rationing of petrol, and its rural situation in the Borders countryside. The new circuits were not only popular with the spectators, but also gave potential Scottish racing drivers access to race meetings on home soil.

On 3 June 1956 a young farmer from Chirnside near Duns, in the Scottish Borders, entered his first race at the old Army base of Stobbs Camp, near Hawick. Using his own Sunbeam Mk3 car he won the over 2 litre sprint; his name was Jim Clark. Clark went on to win two Formula One world championships, but it was while racing at his local circuits, Winfield and Charterhall, that his abilities were nurtured. Clark was probably the greatest motor racing driver of his time, and many would say he was one of the greatest of all time. During his career he won 25 Grand Prix, beating the record of the great Juan Manuel Fangio, and countless other races in other classes. Tragically Jim Clark was killed on 7 April 1969 while competing in a Formula 2 race at Hockenheim in

A well attended race meeting in the 1950s at Charterhall in the Scottish Borders.

Germany. Had he not been killed, how much more might he have achieved?

During 1969 Clark had on three occasions beaten into second place another young Scottish driver, Jackie Stewart. Just as Clark had dominated Grand Prix racing during the 1960s, Stewart dominated the 1970s. Not only did he beat Clark's record of wins, but he went one better with three World Championships. Scotland has produced many other great drivers, such as Ron Flockhart and Innes Ireland, but none have so completely dominated the sport as Jim Clark and Jackie Stewart.

Today there are several young drivers making their mark in motor racing. Many are supported by the Paul Stewart Racing Team run by Jackie's son. But it is not just in single-seat racing cars that Scots have reached the top. Another Borders man John Cleland was the winner of the British Touring Car Championship in 1989 and again in 1995.

The 1951 Le Mans-winning Jaguar at its Edinburgh base in Merchiston Mews. The stylish transporter was built in Falkirk specially for the team.

Scotland has also produced its fair share of top rally drivers. Louise Aitken-Walker from Duns won the ladies' World Rally Championship to become Britain's only female World Champion in motorsport. Andrew Cowan and Jimmie McRae have both reached the top in British rallying, while Jimmie's sons Alastair and Colin are both following in their father's tyretracks. Colin won the 1995 World Rally Championship, driving for Subaru.

Sports-car racing has also had its Scottish exponents. Probably the most successful racing team to come from Scotland was Ecurie Ecosse, which specialized in this class. Founded in Edinburgh in 1951 this small team marked up a remarkable string of successes. The team will probably be best remembered for winning the Le Mans race in 1956 with a Jaguar D-Type, driven by Ron Flockhart from Edinburgh and Ninian Sanderson from Glasgow.

Many well-known drivers raced for the team including the great Jackie Stewart. Sadly the only Scottish cars to be used were a pair of Hillman Imp powered single-seaters, known as Ecosse-Imps, which were raced at the Ingliston Circuit near Edinburgh for the 1966 and 1967 seasons. Many of the team's successes were, however, on Scottish circuits, driven by Scottish drivers.

Ecurie Ecosse did not claim all the honours in the early years; another Scottish team, the Border Reivers, provided stiff opposition. The Reivers name had been used before, and was revived by Ian Scott Watson with the specific aim of giving Jim Clark the opportunity to race a really competitive car. A D-type Jaguar was found and Clark typically won on his first outing. The races at Charterhall during 1958 between these two rival teams must have been a thrill to watch.

Today there are no Scottish racing cars but there is still plenty of talent. The multinational Ligier Formula 1 racing team is run very successfully by Lanark-born Tom Walkinshaw, an ex-racing driver himself. Scotland's latest rising star in Formula 1 is David Coulthard from Twynholm in Dumfriesshire. For 1995 David drove for the Rothmans Williams Renault team and his best results were a win in Portugal and second places in the Brazilian, German and Hungarian Grand Prix. Another World Champion from Scotland?

FURTHER READING

The following books specifically cover the Scottish motor industry:

ALLAN, R J *Geoffrey Rootes' Dream for Linwood*, Minster Lovell 1991.

BROWNING, A S E *Scottish Cars*, Glasgow 1962.

HENSHAW, D & P *Apex, The Inside Story of the Hillman Imp*, Minster Lovell 1990.

OLIVER, G *Motor Trials and Tribulations*, Edinburgh 1993.

WORTHINGTON-WILLIAMS, M *The Scottish Motor Industry*, Aylesbury 1989.

The following contain information on Scottish motor vehicles:

GRIEVES, R *Motoring Memories*, Renfrew 1988.

KIDNER, R W *The First Hundred Road Motors*, South Godstone 1950.

MONTAGU, LORD *Lost Causes of Motoring*, London 1960.

RANSOM, P J G *The Archaeology of the Transport Revolution 1750-1850*, Tadworth 1984.

REES, C *British Specialist Cars*, London 1993.

ROBERTSON, A *Lion Rampant and Winged*, Barassie 1986.

STRATTON, M & COLLINS, P *British Car Factories from 1896*, Godmanstone 1993.

These books are recommended for information on Scottish motorsport:

GAULD, G *Ecurie Ecosse*, Edinburgh 1992.

GAULD, G *Jim Clark The Legend Lives On*, Wellingsborough 1975.

STEWART, J & DYMOCK, E *Jackie Stewart World Champion*, London 1970

PLACES TO VISIT

Museums where Scottish cars can usually be seen on display:

Biggar, Lanarkshire: *Biggar Museums*

Glasgow: *Museum of Transport*

Alford, Aberdeenshire: *Grampian Transport Museum*

Aberlady, East Lothian: *Myreton Motor Museum*

Edinburgh: *National Museums of Scotland*

Other motor museums in Scotland:

Glenluce, Wigtownshire: *Glenluce Motor Museum*

New Lanark, Strathclyde: *New Lanark Motor Museum*

Melrose, Roxburghshire: *Melrose Motor Museum*

Bankfoot, Perthshire: *Highland Motor Heritage*

Elgin, Morayshire: *Moray Motor Museum*

Carse of Cambus, Doune, Perthshire: *Doune Motor Museum*

Scottish vehicles can often be seen at the numerous vintage vehicle events held throughout Scotland during the summer months. This is a small selection:

Glamis Castle, Angus: July

Dunbar, East Lothian: August

Dalmeny House, Midlothian: July

Biggar, Lanarkshire: August

Mellerstain House, Berwickshire: June

Alford, Aberdeenshire: July